Rubber Manufacturing in Malaysia

Rubber Manufacturing in Malaysia

Resource-based Industrialization in Practice

C.C. Goldthorpe

NUS PRESS
SINGAPORE

Published by:

NUS Press
National University of Singapore
AS3-01-02, 3 Arts Link
Singapore 117569

Fax: (65) 6774-0652
E-mail: nusbooks@nus.edu.sg
Website: http://www.nus.edu.sg/nuspress

ISBN 978-9971-69-836-2 (Paper)

National Library Board, Singapore Cataloguing-in-Publication Data

Goldthorpe, Christopher C., author.
Rubber manufacturing in Malaysia: resource-based industrialization in practice/C.C. Goldthorpe. – Singapore: NUS Press, National University of Singapore, [2015]
pages cm
Includes bibliographical references and index.
ISBN: 978-9971-69-836-2 (paper)

1. Rubber industry and trade – Malaysia. 2. Manufacturing industries – Malaysia. 3. Rubber goods – Economic aspects – Malaysia. I. Title.

HD9161.M32
382.45678209595 — dc23 OCN892811711

Cover image: Courtesy of Phil Tomlinson

Printed by: Mainland Press Pte Ltd

Contents

List of Tables

List of Abbreviations

ASEAN	Association of Southeast Asian Nations
BR	polybutadiene rubber
DRC	dry rubber content
EPDM	ethylene propylenediene rubber
EPM	ethylene propylene rubber
ETP	Economic Transformation Programme
EU	European Union
FEER	*Far Eastern Economic Review*
FMM	Federation of Malaysian Manufacturers
GDP	Gross Domestic Product
HICOM	Heavy Industries Corporation of Malaysia
IBRD	International Bank for Reconstruction and Development
IMF	International Monetary Fund
IMP	Industrial Master Plan
IRRDB	International Rubber Research and Development Board
IRSG	International Rubber Study Group
LPG	liquefied petroleum gas
MCA	Malayan Chinese Association
MIC	Malayan Indian Association
MIDA	Malaysian Industrial Development Authority
MITI	Ministry of International Trade and Industry
MRB	Malaysian Rubber Board
MREPC	Malaysian Rubber Export Promotion Council
MRPMA	Malaysian Rubber Products Manufacturers' Association
NDP	National Development Policy
NEAC	National Economic Advisory Council
NEM	New Economic Model
NEP	New Economic Policy
OE	original equipment (tyres)
PAP	People's Action Party

Pernas	*Perbadanan Nasional Berhad*
Petronas	National Petroleum Corporation
PNB	*Permodalan Nasional Berhad*
R&D	research and development
RM	Ringgit Malaysia
RSS	ribbed smoked sheet
SBR	styrene butadiene rubber
SME	small and medium-scale enterprises
SMR	Standard Malaysian Rubber
TARRC	Tun Abdul Razak Research Centre (UK)
TSR	technically specified rubber
UMNO	United Malays National Organization
UNIDO	United Nations Industrial Development Organization
XNBR	nitrile latex, acrylonitrile butadiene latex
XSBR	SBR latex, carboxylated styrene butadiene latex

Acknowledgements

This book is the outcome of a research study carried out at the Bradford Centre for International Development at the University of Bradford under the supervision of Prof. John W. Cusworth. Additional guidance during the course of the study was provided by Prof. Tom R. Franks. The comments of Prof. Colin Kirkpatrick and Dr David J. Potts have contributed positively to the end result.

Loh Ai Tee was good enough to read through the manuscript and gave advice on each chapter based on her long experience with the Asian Development Bank. Choo Weng Choong provided insights on the chapters dealing with industrialization policy. Michael H. Mellor and John F. McEwen gave comments on the technical aspects of rubber production and processing. In Kuala Lumpur, Michael Chang assisted with the use of office facilities during the administration of the questionnaire. The text was edited by Chang Wei-ling to her usual exacting standards. To all these individuals, my sincere thanks.

In this project help and support have been given by many persons. My thanks to: Colin Barlow, Henry S. Barlow, A.F.S. Budiman, M.R. Chandran, Axèle Giroud, Graham Howarth, Peter Jenkins, W. Tissa Perera, N. Thirugnanasothy, Wan Abdul Rahaman Wan Yaacob, Philip S. Watson and Frank A. Wilson.

At NUS Press, the encouragement of Paul Kratoska has proved invaluable. Thanks are also due to Joann Keong and Lena Qua for preparing the manuscript for publication.

The work is dedicated to my wife, Rosa Chang, for her love and friendship over many years of married life.

Map of Peninsular Malaysia

CHAPTER 1

Development through Industrialization

Malaysia: Commodity Producer to Industrialized Economy

At Independence in 1957, Malaysia was a classic colonial economy producing two industrial raw materials, rubber and tin, for export on international commodity markets. During the 1950s the country was the world's largest producer of natural rubber, and the agricultural sector accounted for 40 per cent of Gross Domestic Product (GDP) compared to 6 per cent for the manufacturing sector. Rubber estates and smallholdings occupied about 65 per cent of the entire cultivated area of Peninsular Malaysia and the crop provided some 60 per cent of export earnings. Today, the economy is broadly based and diversified, with the manufacturing sector accounting for 30 per cent of GDP and the country classed as a newly industrialized economy.

In 2008, 50 years after Independence, Malaysia had become a major producer of manufactured goods that contributed over 50 per cent by value of total exports. In the commodity and commodity-based sectors, tin deposits basically had been worked out while exports of crude petroleum oil and liquefied natural gas; palm oil and palm oil products; and timber and timber-based products had eclipsed natural rubber as export earners, as shown in Table 1.1. The contribution the rubber manufacturing industry made to export revenues is not detailed in Table 1.1 because the manufacture of rubber products was classed in government statistics, at the time, under the "other manufactured products" sector which provided 13 per cent of export revenue. In manufacturing, figures for the value of exports of manufactured products in 2010 demonstrate the importance of the electronic and electrical sector at 51 per cent, while showing the revenue from rubber products at only 3.2 per cent by value. This figure was less than the exports of other manufacturing sectors such as chemical and petroleum products, and machinery and equipment. Overall, exports of raw rubber and rubber products did not feature strongly in the country's economy.

Table 1.1 Major Exports: 2008

Product	*Per cent*
Electrical and electronic products	38.5
Other manufactured products	13.0
Petroleum and LNG	17.0
Palm oil and palm products	10.0
Timber and timber products	3.5
Rubber	1.2

Source: Malaysia: Ministry of Finance 2009.

The importance of the electrical and electronic products sector to the national economy is such that it has attracted a number of studies by both Malaysian and foreign academics. Similarly, in agriculture, the growth of the oil palm plantation industry and associated palm products industry since the 1950s has been of interest to economists in recent years. Gopal (1999), for example, analyzed the development of the palm oil refining sector which, by the 1990s, processed 99 per cent of the domestic production of crude palm oil. The refining of crude palm oil extracted from the fruit of the oil palm (*Elaeis guineensis*) in large oleo-chemical processing plants uses advanced engineering technology and adds high value to the crude product. The rapid development of this resource-based industry from 1970 onwards is regarded as a success in Malaysia's industrialization programme (UNCTAD 1998; Gopal 1999; Rasiah & Jomo 1999; Jomo 2001; Rock & Sheridan 2007). On the other hand, there is no similar analysis of the group of industries that uses natural rubber, a product of the tree crop, *Hevea brasiliensis*, to manufacture products as diverse as tyres, medical examination gloves and gaskets. The present study, therefore, is an attempt to fill that gap in economic studies on the development of the manufacturing industrial sector in Malaysia and, more particularly, on industrialization based on the utilization of an agricultural resource grown in the country.

By the second decade of the 21st century, Malaysia had progressed from commodity producer to a country with a substantial manufacturing base, with the industrial sector contributing approximately one-third of Gross Domestic Product. The purpose of the study is to assess the contribution the rubber industrial sector has made to the process of industrialization. It does this by tracing the development of the rubber products manufacturing industry from the 1920s to the present day. In doing so it examines the role the rubber manufacturing sector has made to the transformation of the Malaysian economy from primary commodity producer to a newly industrialized country. It is argued that the production of rubber goods adds value to locally grown natural rubber through the manufacture of rubber articles in an export-oriented industry. Rubber manufacturing, it is suggested, has played its part in Malaysia's industrialization programme and continues to have an important role in developing the country's resource-based industrial sector. The rubber industry taken as a whole is an example of a vertically integrated resource-based industry with domestic production of natural rubber used as a raw material to produce a range of manufactured goods for export markets. The study argues that the example of Malaysia in utilizing the products of the tree crops, rubber and oil palm, to manufacture consumer and intermediate products for sale on international markets may provide lessons for other tropical agricultural commodity exporters in their quest for economic development.

Porter Model of Industrialization

One of the planks of international trade theory is that developing countries are able to increase exports by following policies to exploit their comparative advantage. At the early stages of development, tropical countries grow agricultural crops adapted to the local climatic and ecological environment to enter the international trade in commodities. Primary crop producers, at a later stage, have the opportunity to move up the ladder of comparative advantage by developing a secondary manufacturing industrial base that utilizes the crop as raw material. Transforming the agricultural commodity into manufactured products that are sold into export markets adds value to the commodity and increases foreign exchange revenues that may be used to finance national development projects. Theorists belonging to the neo-classical school argue that an export-oriented industrialization programme allows a country to develop more rapidly than when protectionist policies are adopted to support domestic manufacturing through an import-substitution programme. Contemporary trade theory advances the argument that governments can promote rapid economic development by implementing policies to build on the comparative advantage of abundant natural resources by creating groups of competing industries to manufacture technologically complex products from basic resources.

The thesis that the process of industrialization may be enhanced by the establishment of groupings of high-tech industries utilizing domestically produced raw materials is particularly associated with Porter (1990). In *The Competitive Advantage of Nations*, M.E. Porter classified production factors into a number of broad categories: physical resources, knowledge resources, capital resources and infrastructure. He then divides the factors into two classes, basic and advanced factors. Basic factors are passively inherited or their creation requires only modest investment; they include natural resources, climate, geographical location, and an abundant supply of unskilled and semi-skilled workers. A nation's advanced factors are often built upon the basic factors and comprise items such as modern communication systems, high quality and robust infrastructure, high investment in human resources, a specialized and well educated labour force, and well endowed research institutions and other knowledge resources.

The Porter model recognizes that whilst the advantages associated with basic factors have a considerable effect on the ability to compete, these advantages can easily be eroded by a lack of advanced factors. An industry is most likely to be competitive where there is an open economy with producers integrated into a discerning market, where there are supporting industries, and there is rivalry among companies for market share. Porter argued that the central task of government trade policy in less developed countries is to encourage movement from comparative advantage dependent on natural resources and other basic factors to the creation of advanced factors. He saw intense competition among clusters of rival firms both domestically and in foreign markets as being the main source of economic benefits in a dynamic trade model.

Structure of the Book

The book provides an economic history and examination of the rubber products manufacturing industry in Malaysia that contribute to the study of resource-based industrialization in the field of development studies. As a case study, it is a practical example of a successful industrial programme that augments the existing literature on the process of industrialization in emerging economies. It adds to the substantial volume of studies on the transformation of Malaysia from an exporter of rubber and tin to a newly industrialized country over the past 50 years and more, and the role the Malaysian government has taken in promoting industrial development. This study on rubber is a useful addition to the literature on the success of Malaysia's resource-based industrialization programme as the published material hitherto has focused on transforming locally produced palm oil into processed products by a newly established oleo-chemical industry.

Chapter 2 provides background information on the world rubber industry and the technology of rubber manufacture so that later chapters that describe the Malaysian manufacturing sector may be placed in context. The chapter discusses the importance of rubber as an industrial raw material and presents an overview of the development of the world rubber industry before focusing on international production and trade in natural rubber. The processes involved in converting field latex into forms suitable for use by manufacturing industry are then described. The manufacture of rubber products is technologically driven and the discussion reviews the major scientific discoveries that have enabled natural rubber to become an essential component in a range of industrial and consumer products. The chapter goes on to describe the basic manufacturing processes used in the three main product sectors in the rubber manufacturing industry, namely the fabrication of tyres, the production of general rubber goods and the manufacture of latex products.

An account of the political economy of Malaysia is the subject of Chapter 3. The account chronicles the country's progress from primary producer to an industrialized economy. The discussion emphasizes the important role the agricultural sector took in the colonial period and the part it played in the development plans of post-Independence Malaysia. The significant influence the New Economic Policy has had since 1970 on the political process as well as economic policy and strategy is discussed.

Chapters 4 and 5 examine, in turn, the development of the industrial sector in the economy of Malaysia and the rubber manufacturing sector in particular. The policies adopted by the authorities to encourage the development of a manufacturing sector from the time of colonial rule and Independence through the post-Independence years are traced in Chapter 4. The focus of Chapter 5 is the part natural rubber plays in the economy and includes an account of the manufacturing industries that make rubber products from natural rubber. Basic economic data of production, export and import figures for the rubber production sector are described. Rubber manufacture is investigated from the establishment of the first factories in the 1920s to 2005 at the conclusion of the Second Industrial

Master Plan. A more wide-ranging examination follows the contribution the rubber production and manufacturing sectors have made to the national economy over the period 2005 to 2012. The chapter concludes with a discussion on linkage effects with other sectors of the economy and makes the observation that the industry has a dualistic structure.

Chapter 6 provides an overview of the rubber manufacturing industries and classifies the industries into eight production sectors based on the technology employed and the end use of the products. The chapter continues with an analysis of each of the product sectors in turn, based on information published in trade directories. A detailed examination of the automotive, engineering and industrial product sector that manufactures intermediate parts and components for other industries is undertaken in Chapter 7. The analysis considers the range of products made by this sector, the structure of the sector, export sales, forward and backward links with other industries and the sourcing of industrial technology. Chapter 8 describes the tyre, latex and low-tech product sectors and limits the analysis to an examination of industrial structure and the pattern of export sales.

The topics of structural dualism and technology transfer are considered in detail in Chapter 9 which also summarizes the results of the analysis for the entire rubber manufacturing industry. Chapter 10 considers the wider policy implications of the research study and discusses the part the rubber industrial sector plays in the context of contemporary ideas in development economics and international trade theory. The overall conclusion is that the rubber manufacturing industries are well integrated into the economy and have contributed positively to the success of Malaysia's development to become an industrialized economy.

CHAPTER 2

Rubber: An Essential Industrial Raw Material

End Uses of Natural and Synthetic Rubber

There are two sources of rubber: natural rubber and synthetic rubber. Natural rubber is an agricultural commodity produced from the sap or latex of the tree, *Hevea brasiliensis*, a species indigenous to the Amazon basin. Synthetic rubber, on the other hand, describes a diverse group of over 200 related chemical compounds made in large-scale chemical engineering plants from petroleum oil in countries with a well developed petroleum refining industry. Both types of rubber are traded on the world market and used in the manufacture in a range of consumer and intermediate products.

Rubber, derived either from tapping the bark of the *Hevea* tree or from petroleum feedstock, is a major industrial raw material used in the manufacture of domestic and industrial products essential to modern life. Table 2.1 provides a list of many common articles made from this versatile substance.

The market for rubber manufactures is divided into three major sectors: tyres, general rubber goods and latex products. Tyres and general rubber products are produced from solid rubber, whereas liquid latex concentrate is the raw material for the manufacture of products such as household and medical examination gloves. The general rubber sector consists of a number of subdivisions such as engineering and industrial products, footwear, adhesives and other general merchandise. Figures for the uptake of rubber by end use provided by the International Rubber Research and Development Board (IRRDB 2007) are shown in Table 2.2.

Table 2.1 Rubber Manufactured Items

Adhesives	Hoses
Auto-components (bumpers, engine mounts)	Hot water bottles
Balloons	Mats
Bridge bearings	Milking machine liners
Carpet underlays	Printing rollers
Condoms	Roofing material
Conveyor belts	Seals and gaskets
Diving suits	Sheeting
Elastic bands	Shoes and sandals
Elastic thread	Tank liners
Electrical insulation	Toys and balls
Foam mattresses	Transmission belts
Fenders	Tyres and inner tubes
Gloves (household)	Walk surfaces
Gloves (industrial)	Weatherstrip
Gloves (medical)	Windscreen wipers
Gum boots and waders	

Source: Adapted from Barlow et al. 1994, Table 1.1.

The tyre sector is by far the single largest user of rubber with almost 70 per cent of consumption although more recent figures produced by the International Rubber Study Group (2013) indicate a fall in uptake by the tyre industry to 60 per cent.

The mass manufacture of motor vehicles from the early 20th century to the present day stimulated and continues to influence the growth of the rubber industry. Production levels of new vehicles create demand for original equipment tyres, while the number of passenger cars and commercial vehicles on the road determines the market for replacement equipment. Motor vehicles, railways and aircraft make up the wider automotive

Table 2.2 End Uses of Rubber

Products	*Per cent*
Tyres	68
Latex products	8
Engineering/Industrial	8
Footwear	5
Adhesives	3
Other products	8

Source: IRRDB 2007.

sector for the transportation of goods and people. This sector, besides the demand for tyres, is also a significant market for non-tyre products including transmission belts, seals, gaskets and hoses as engine parts, and weatherstrip and wipers for doors and windows. It is estimated that the uptake of rubber by the automotive manufacturing industry as a whole is in the region of 75 per cent of total rubber consumption.

Development of the World Rubber Industry

The history of the discovery of natural rubber in tropical South America by European explorers and its subsequent introduction as a cultivated crop into the Far East and West Africa is well documented in the literature (Allen & Donnithorne 1954; Drabble 1973; Voon 1976; Barlow 1978; Coates 1987; Dean 1987; Baulkwill 1989; Jones & Allen 1992; Barlow et al. 1994; Tate 1996; Loadman 2005; IRRDB 2006; Jackson 2008). In the 19th century, rubber was collected from wild trees growing in the Amazonian regions of Brazil, Bolivia and Peru, and exported to industrial centres in Europe and the north eastern states of the USA for the manufacture of a range of items, including waterproof clothing and footwear, conveyor belts and insulation for electric cables. The transfer of the *Hevea* rubber tree to Asia from its native habitat owes as much to British imperial, economic expansion in the Far East during the late 19th century as it does to growing demand from early industrialists. Sir Clements Markham, an official in the India Office, arranged expeditions to the Amazon to collect seeds for transfer to British colonies in Asia. In 1876, a British adventurer, Henry Wickham shipped a consignment of seeds, gathered around the River Tapajos near the town of Santarem on the middle reaches of the Amazon in Brazil, to Kew Gardens where over 2,000 were germinated and the seedlings successfully transferred to the Botanical Gardens in Ceylon. The cultivation of natural rubber as an agricultural commodity crop, grown on large plantations and smallholdings, expanded across the humid tropics from the Wickham collection. The total area planted with rubber trees in the 1990s was calculated to be over nine million hectares (IRSG 1990). Small farmers, typically with holdings of two hectares or less, cultivate about 85 per cent of the rubber area. The remaining rubber is grown on estates owned, on the whole, by large corporate plantation companies. It is estimated that over 20 million families in tropical developing countries are dependent on rubber cultivation for their livelihood as either estate workers or small growers (IRRDB 2007).

The first practical pneumatic tyre for bicycles was invented in 1888 by a Scottish veterinarian, John Boyd Dunlop who practised in Belfast, Northern Ireland. This created a new demand for rubber as cycling became popular and the wealthy classes adopted the horseless carriage as a means of transport. The growth in demand for rubber as a raw material for tyre production increased exponentially in the early years of the 20th century with the development of the mass production of motor vehicles, including passenger cars, trucks and buses. It was this demand that stimulated the flotation of plantation companies on the stock exchanges of London, Paris and Amsterdam for the establishment of rubber estates in present day Sri Lanka, Malaysia, Cambodia, Vietnam and Indonesia. Restrictions on rubber production imposed by the colonial powers during economic recessions in the 1920s and 1930s caused American tyre companies to establish their own rubber plantations in the Netherlands East Indies, Liberia and Brazil in order to safeguard supplies. It also had the effect of encouraging farmers to plant rubber on their smallholdings in southern Thailand.

The Japanese invasion of Southeast Asia in 1941 cut off the supply of a strategic war material to tyre manufacturers and producers of rubber goods in Western countries. The disruption to supplies of raw natural rubber was the catalyst for the establishment and expansion in the USA of the synthetic rubber industry that uses petroleum oil feedstock and was based on German technology developed in the 1914–18 World War. After 1950, increased demand, mainly from the expanding passenger car and commercial vehicle industries, led to the planting of new rubber growing areas where climatic conditions are suitable for *Hevea* production, notably in the southern regions of Thailand, India and China in Asia, and the Ivory Coast in West Africa. In more recent years, the continued strong demand for rubber has stimulated the development of new rubber plantings in Indochina and countries in western and central Africa.

World Production and Consumption of Natural Rubber

The growth in world demand for natural rubber over the past hundred years is illustrated in the following figures: production was 45,000 metric tons in 1900; 1.9 million metric tons in 1950; 3.3 million metric tons in 1975 and 6.8 million metric tons in 2000. In 2010 and 2012, total world output was 10.4 and 11.3 million metric tons (Table 2.3) respectively, with production concentrated in Thailand, Indonesia, Malaysia and Vietnam. These four Southeast Asian countries account for approximately 80 per cent of global production. In round figures, Thailand produces almost four million metric tons, Indonesia three million metric tons, Malaysia and Vietnam each one million metric tons per year. Other important producers in Asia are India and China which are both major importers of raw rubber for their expanding industrial sectors. Countries in South America and Africa are relatively unimportant rubber producers compared to countries in Asia.

Table 2.3 World Natural Rubber Production: 2010 and 2012

	000 metric tons	*000 metric tons*
	2010	*2012*
Asia	9,662	10,521
Africa	464	494
South America	267	312
World total	10,393	11,327

Source: IRSG 2013 (includes balancing adjustments).

Table 2.4 World Natural Rubber Consumption: 2010 and 2012

	000 metric tons	*000 metric tons*	
	2010	*2012*	
Asia/Pacific	7,565	7,879	
European Union	1,132	1,072	
Other Europe	402	388	
North America	1,071	1,666	(2012 figures includes
South America	616	na	South America)
Africa	101	na	
World total	10,807	11,005	

Source: IRSG 2013 (includes balancing adjustments).

The consumption of natural rubber by manufacturing industry in 2010 was 10.8 million metric tons and 11 million metric tons in 2012 (Table 2.4). The most important geographical area for rubber manufacture is Asia, accounting for approximately 70 per cent of global natural rubber consumption with China, Japan and India being the largest producers of rubber goods. In the West, important rubber manufacturing centres are the USA and the 28 countries of the European Union.

Technology of Rubber Processing

Field latex with a dry rubber content (DRC) of approximately 35 per cent rubber solids is unstable and has to be processed into material that can be utilized in industrial operations. There are two major types of raw material used in rubber manufacturing industries. These are solid rubber that goes into the production of tyres and general rubber goods, and liquid latex concentrate which is used for the manufacture of latex products. A major aim of processing is to convert naturally produced *Hevea* rubber of a heterogeneous nature, derived from a variety of sources, into a standard, industrial raw material of guaranteed quality that can be used in automated factory operations.

Wild rubber in the Amazon basin was processed by individual tappers who collected latex from scattered *Hevea* trees in the native rain forest. A wooden paddle was dipped into the latex and rotated over a smoky fire to produce balls of "cured" rubber for the export trade. The method of preserving solid rubber by smoking was adapted by large-scale estates with the development of the plantation rubber industry in the Far East. Smoked rubber was sold to visual grading specifications set by the rubber commodity trade and packed into large bales that required cutting before use in manufacturing machinery. Synthetic rubbers, on the other hand, were marketed according to technical specifications in smaller, easily handled, wrapped blocks so that after 1945 synthetic rubber greatly increased its market share at the expense of naturally grown rubber. The natural rubber industry reacted by introducing new methods of processing, grading and packing to compete with producers of the synthetic material in response to demand from manufacturers of rubber goods.

Technically Specified Rubber

Rubber produced on estates and smallholdings is processed in central factories into Technically Specified Rubber (TSR) with guaranteed technical specifications. Each of the major producing countries has its own approved scheme to guarantee the quality of its processed rubber. Primary processing under TSR schemes typically produce five basic grades of dry rubber based on dirt content and whether the raw material is tree latex or field coagulum (cup lump and tree lace). The basic steps in producing technically specified rubbers are: bulking and blending; coagulation of tree latex; size reduction; granulation; drying and baling. There are differences in technique, depending on the method and machinery employed.

In all processes, field latex is bulked in a reception tank before passing into coagulating tanks where controlled coagulation takes place by the addition of formic acid. Partition plates are inserted into the coagulation tank in order to divide the coagulum into slabs of spongy material. When coagulation is complete, the vertical partitions are removed and the tank is flooded with water for the slabs of coagulum to be floated towards the processing machinery. In the *Heveacrumb* process, coagulated latex slabs are passed through differential rotating rollers of a creping machine to produce thin crepe with a lace-like texture. Field coagulum is cleaned by slicing in a heavy grooved macerator, followed by passing through a creping mill battery where further cleaning and blending take place. Crepes derived from latex and field coagulum with the addition of a small amount of castor oil are fed into a hammer mill or granulator that converts the crepe material into crumbs or granules of rubber.

The *Dynat* method involves feeding latex coagulum and field coagulum material through a rotary cutter for initial size reduction before granulation takes place by a process of extrusion and cutting. Granulated rubber in both the *Heveacrumb* and *Dynat* methods is transferred by mechanical conveyors through a hot air dryer. After leaving the dryer, the rubber is baled in a hydraulic press, and then each bale is sealed in a plastic wrapper bag to prevent contamination from extraneous dirt (Morris 1989: 472–83; Kuriakose 1992: 386–91; Barlow et al. 1994: 150–2, MRB 2007).

Latex Concentrate

The centrifuge method is used for the production of latex concentrate in a batch process in custom-built latex factories. Field latex is collected immediately after tapping and treated by the addition of ammonia gas at a collection station to inhibit auto-coagulation caused by bacterial action. Preserved latex is transported by road tanker to the latex factory within the space of a few hours where it is discharged into reception tanks in order for blending to take place. Bulked latex is fed into rows of industrial centrifuges that are started and made to run continuously until the dry rubber content of the latex reaches 60 per cent. The latex concentrate is run off into storage tanks fitted with mechanical stirrers. When the tank is full, the contents are stirred for up to 24 hours before the concentrate is transported locally to latex goods factories or to ports for shipping (Edgar 1958: 470–90; Morris 1989: 484–93; Kuriakose 1992: 375–81).

Ribbed Smoked Sheet

Before the introduction of the technically specified rubber schemes almost all tree latex was made into ribbed smoked sheets (RSS), whereas field coagulum and small amounts of latex were processed into crepe rubbers. In RSS production, the coagulum is passed from the coagulating tank through a sheeting mill in estate-scale production or through hand mangles on smallholdings. A sheeting battery typically consists of six pairs of rollers with increasing nip, the last pair being grooved to give a ribbed pattern to the wet coagulum. The ribbed design increases the surface area of the coagulum for faster drying in the smokehouse. It also prevents the dried sheets from sticking together when they are stacked in the packing shed before baling. After passing through the sheeting battery, the coagulum is cut into sheets that are hung on trolleys to drip-dry before the trolleys are moved into a smokehouse or tunnel dryer. The drying process usually takes four days to complete. The RSS is then taken into a packing shed where grading is carried out before the sheet rubber is pressed into bales (Edgar 1958: 360–430; Morris 1989: 462–6; Kuriakose 1992: 382–6; Barlow et al. 1994: 152–3).

Crepe Rubber

The preparation of crepe rubbers involves passing coagulum through a creping battery consisting of a series of pairs of driven rollers that rotate at different speeds. Pale crepe grades are produced from coagulated field latex and the skim rubber residue from the latex centrifugation process: the material first passes through diamond-shaped, grooved rollers and the final set of rollers is smooth. Pale crepes are still made by a small number of processors for specialist applications, such as medical equipment and shoe soles for fashion footwear where high purity and a light colour are important (Edgar 1958: 431–59; Morris 1989: 467–72; Kuriakose 1992: 391–3).

Rubber Polymer Science

The science of the manufacture of rubber into finished products is studied within the discipline of rubber polymer science, an applied branch of polymer chemistry. There is an extensive body of literature on polymer materials science, rubber science and technology, and rubber industry production methods that is highly specialist in nature. A standard text, for example, is *Natural Rubber Science and Technology* (Roberts 1988), although, a more general account of rubber polymer chemistry is given by Loadman (2005).

The terms *rubber* and *elastomer* are used to describe a group of materials that have the property of elasticity. An elastomer is defined as a substance that will return rapidly to its original size and shape after substantial deformation by stress and release of the stress. Thus, a rubber band can be stretched many times its length without breaking but will revert quickly to its original measurement when the stretching force is disconnected. The property of elasticity is dependent on the fact that all elastomers consist of long flexible molecules known as linear high polymers. Strength and flexibility are imparted to natural rubber and other elastomers during manufacturing when the polymer molecular structure is chemically modified through the formation of cross-links by the use of fillers and the process of vulcanization. It is this combination of elasticity and flexibility, coupled with strength that gives natural and synthetic rubbers their unique properties. This also makes rubber an essential raw material in a range of industrial processes.

Technological Developments in Rubber Manufacturing

The account earlier has shown that demand, principally by tyre manufacturers, was a key factor in the expansion of rubber tree crop cultivation. However, advances in rubber polymer science and industrial technology were also of vital importance in the development of rubber manufacturing industries for an expanding portfolio of commercial applications. The major technological discoveries that helped increase the industrial uses of natural rubber are therefore described in this section.

From the late 18th century onwards, natural scientists, particularly in France, worked on ways of making waterproof coats and the production of elastic thread for fashion garments by experimenting with organic solvents such as turpentine, and dipping yarn and cloth in the resultant rubber solutions. The rubberized textiles produced by this method were, however, brittle in winter and became sticky in warm weather. In 1823, the Glasgow industrialist, Charles Macintosh developed a process of waterproofing cloth by dissolving rubber in coal tar naphtha and applying the solution between two sheets of fabric. Macintosh's method allowed the large-scale production of raincoats and other waterproof fabrics because the textile layers protected the rubber, and the solvent was a cheap waste product of the expanding town gas industry.

The next significant steps took place, between 1820 and 1845, with the invention of three types of machinery capable of processing rubber in its solid state. In 1820, the English inventor, Thomas Hancock developed a machine that transformed hard, solid rubber to a soft, plastic material by shearing it between two rotors enclosed in an iron vessel. This process known by the technical term of *mastication* enables softened rubber to be formed into shapes, such as boots, by moulding. The *calender* machine was patented in the USA by Edwin Chaffee in 1836. Calendering produces sheets of rubber of uniform thickness and makes possible the bonding of rubber to canvas for the manufacture of belting. The *extruder* machine was invented to make insulated cables for telegraphic communications in 1845, and was adapted later to produce electric cables and wiring.

Vulcanization or *curing* is the term for the chemical process during which sulphur combines with rubber by the application of heat. The process imparts strength and flexibility to rubber when cross-links are made between sulphur atoms and the long, polymer molecules of natural rubber to form a complex, three-dimensional, matrix molecular structure. Vulcanized or cured rubber retains the property of elasticity and is stable under heating and cooling, unlike non-vulcanized rubber that is soft and sticky at high temperatures but hard and brittle at low temperatures. The discovery of vulcanization is a matter of contention but is jointly attributed to Charles Goodyear in the USA and Thomas Hancock in England. The American, Goodyear, is generally acknowledged as the first in 1839, with an accidental discovery on a hot stove. However, the facts are that Hancock obtained a British provisional patent for the process in November 1843 and a final patent in May 1844, whereas Goodyear received his US patent in June 1844. As the polymer scientist, John Loadman (2005: 64) commented, "Whilst Goodyear may well have been the first person to vulcanise rubber, he certainly had no control of his process at the time of Hancock's patent, when the latter was able to illustrate his complete understanding and control of the chemistry."

The development of machinery able to deal with solid rubber together with the process of vulcanization greatly accelerated the evolution of new rubber products. There is general consensus that the empirical discoveries of Goodyear and Hancock established the foundation of the rubber products manufacturing industries of today. Barlow et al. (1994: 190), for example, regarded vulcanization as "the crucial technological breakthrough enabling rubber to become a pre-eminent industrial material". The viewpoint of the International Rubber Research and Development Board (IRRDB) is that "the twin developments of mastication by Hancock and vulcanization by Goodyear enabled the birth of the modern rubber industry" (IRRDB 2007). From the 1850s onwards, rubber was being used to make industrial and consumer goods, such as springs for railway rolling stock, conveyors for grain elevators and mines, electrical and telegraphic cabling, waterproof fabrics and rubber footwear.

The assembly line method of automobile production developed in the USA by Henry Ford created a mass market for motor cars. Large-scale motor vehicle manufacturing placed pressure on the tyre industry to make a product that was long lasting and affordable by car owners of moderate financial means. Two developments, the use of carbon black filler in 1904 and the introduction of the Banbury mixer in manufacturing operations in 1916, are regarded as significant in satisfying these needs and contributing to the growth of the tyre industry. The identification of carbon black, produced by the incomplete combustion of natural gas, greatly improves the wear-resistance and tensile strength of natural rubber. The Banbury internal mixer enables large quantities of rubber to be masticated and mixed with carbon black, vulcanizing materials and other chemicals in a relatively short time. A major advance in the 1930s was the discovery that some mineral oils can be incorporated into rubber mixtures, making the rubber easier to work during manufacture.

A new industry within the rubber manufacturing sector came into being in the late 1920s with the introduction of liquid latex concentrate. Products manufactured from latex concentrate replaced existing goods produced from rubber solutions because of their superior properties, cheaper production costs and ease of manufacture. Latex concentrate allows the inexpensive, large-scale production of dipped goods, extruded thread, foam rubber and adhesives. Dipped products, such as surgical gloves and condoms, are of higher quality compared to similar products made by the solution process. It became possible for clothing manufacturers to design new forms of outer garments and underwear incorporating latex extruded thread, while foam rubber is used in the furniture and bedding industries.

The large-scale production of synthetic rubbers that took place in the 1940s was used as an alternative material to natural rubber because supplies of *Hevea* rubber from Southeast Asia were cut off during the Second World War. Similarly, during the Cold War, the Soviet Union developed massive capacity to produce synthetic rubbers for strategic reasons. Low cost, "general purpose" synthetic rubber with a molecular structure similar to *Hevea* rubber was a direct competitor to natural rubber, particularly in the tyre industry, and the natural product lost ground to synthetic materials. However, the competitive position of *Hevea* rubber changed in the 1970s with increases in the cost of crude petroleum and improvements to the quality of natural rubber by the introduction of technically specified rubber. At the same time, advances in polymer science that led to greater understanding of the differences in properties between synthetic and natural rubbers prompted manufacturers to use blends of both materials. Today, synthetic rubber and natural rubber are generally regarded as complementary products rather than competitors, each with its own cost advantage and unique technical properties suitable for a particular purpose. Both elastomers are used in blends for the majority of manufactured articles, the relative quantities of each rubber being dependent on the desired technical specification of the end product. The consumption of synthetic and natural rubbers by manufacturing industries on a global basis has stabilized for many years at an approximate 60:40 ratio in favour of synthetic materials. In 2012, for example, total world rubber consumption was 26 million metric tons of which synthetics accounted for 15 million and natural rubber 11 million metric tons respectively (IRSG 2013).

Basic Processes in Manufacturing

Providing a non-technical description of the basic processes involved in the manufacture of rubber goods for the general reader concludes this chapter.

The purpose of vulcanization is to modify the molecular structure of the rubber polymer through the formation of cross-links with vulcanizing chemicals by the application of heat. The most common vulcanizing agent is sulphur which is used with small amounts of other materials, including zinc oxide and organic compounds such as stearic acid that activate the process. These chemicals are used to accelerate the curing process, assist and improve manufacturing operations, provide strength, add bulk, reduce cost, colour the rubber and impart specific properties to the end product, such as resistance to ageing. In the rubber industry, the mixing of different substances such as fillers, accelerators and curing agents with raw rubber during mastication is referred to as *compounding* and the generic term for all these materials is *compounding ingredients.*

Compounding rubber is a complicated operation because all the materials interact to determine the properties of the final product so that mix formulation is a skilled procedure within the overall manufacturing process, especially when production cost factors have to be taken into consideration. The objectives of compounding have been summarized by Crowther et al. (1988: 177) thus:

- to facilitate processing and fabrication;
- to ensure a rapid throughput with minimal reject rate;
- to achieve the required balance in vulcanizate properties; and to provide durability.

Compounding solid rubber requires large-scale mixing machinery for economic throughput and, in the case of latex compounding, special equipment to handle a liquid product. Furthermore, cost savings can be made when items such as accelerators for vulcanization and carbon black filler are purchased in bulk. Specialist compounding companies known as compounders produce compound rubber mixes or, *masterbatches* of natural and synthetic rubbers with oil, carbon black and other chemicals that are sold to small and medium-sized manufacturers. The masterbatch is added to raw rubber and other ingredients during manufacturing in order to facilitate factory operations.

The methods used in the manufacture of products from dry rubber produced in bales on the one hand, and liquid latex concentrate on the other, are so different that latex and dry rubber manufacturing are regarded as two separate sectors of the rubber industry. Similarly, tyre production, the largest consumer of natural and synthetic rubbers, employs a specialized technology in the dry rubber sector so that it too is regarded as a separate industry. The following sections examine, in turn, the manufacturing technologies for tyres, general rubber products and latex goods.

Tyre Manufacture

The tyre industry is dominated by a small number of transnational corporations that operate on a worldwide basis: the three largest multinational firms are Goodyear (USA), Michelin (France) and Bridgestone (Japan). There are also smaller manufacturers that are strong as specialist tyre producers, as are companies that produce tyres for regional markets. The market for tyres is divided into two categories: *original equipment* (OE) tyres fitted to new vehicles at the time of manufacture, and *replacement* tyres purchased to replace OE tyres when the treads are worn out. *Retreading* worn-out tyres is important in extending the life of heavy commercial vehicle and aircraft tyres in the road haulage and aviation industries.

Pneumatic tyres are complex engineering products manufactured to rigorous technical specifications. Tyre factories are typically large-scale operations with highly automated production lines because economies of scale offer cost advantages through long production runs. Automation also allows high quality outputs to be obtained with minimum inputs and technical rejects.

The basic design of a tyre consists of a casing (*carcass*) made out of layers of rubberized fabric and steel cord (*plies*) attached to steel wires (*beads*) embedded on the two sides of the carcass. The function of the steel beads is to attach the tyre to the rim of the wheel. The carcass has an overlay of *sidewalls* that extend from the beads to the tyre *tread*. The tread is built on a number of *tread bracing layers* or *breakers*, and it is that part of the tyre that comes in contact with the road surface. Blends of natural and synthetic rubbers are used in the carcass, sidewalls and treads, the amount of each rubber is dependent upon the tyre component and the type of tyre, for example, for motor car, heavy truck or agricultural tractor. Tyres are fabricated or built from the individual components on an open-ended metal drum shell mounted on a rotating shaft. After fabrication, the raw tyres are transferred into a compression mould that gives the required shape and tread pattern. Vulcanization then takes place by applying heat from an external source. After cooling, the finished tyre products undergo quality control testing procedures.

General Products Manufacture

There are three main stages in the fabrication of general rubber goods:

- Mixing
- Shaping
- Vulcanization

Mixing

Mixing is carried out in heavy-duty, internal mixers with a capacity to process batches of 200 kilograms of rubber in two minutes. This operation has two functions: mastication and compounding. Firstly, the solid rubber is softened in the process of mastication by shearing between rotary mixers or mills. Secondly, the rubber is admixed with compounding ingredients that typically include masterbatch, additional fillers such as carbon black or silica, vulcanizing components, accelerator chemicals and protective agents. In many manufacturing operations, it is the practice to add synthetic rubbers during the mixing process to produce a blend of natural and synthetic materials in order to combine the properties of each elastomer. Mixing operations generate considerable heat and the resultant compounded rubber mixture is soft and plastic, therefore easily shaped.

Shaping

The soft rubber mix is formed into the final product by three common shaping techniques: *moulding* under pressure; *extrusion* through a die; and *calendering* between rollers.

Vulcanization

The process of vulcanization is part and parcel of the manufacturing operation in the production of moulded articles when curing takes place at the same time as shaping in heated moulds. Extruded and calendered products are vulcanized after shaping.

Latex Goods Manufacture

The production of articles from latex concentrate, made by centrifuging raw latex from *Hevea* trees, is a separate manufacturing sector from the industrial production of tyres and general rubber goods. This is because the machinery used, factory techniques employed and marketing structure for finished products differ greatly from the rest of the industry. Articles may be manufactured from pure natural rubber latex, a mixture of natural latex and synthetic latex, or synthetic latex. The manufacture of latex goods involves both compounding and vulcanization similar to dry rubber manufacturing. Larger producers generally operate fully integrated processing plants but many smaller processors purchase latex compounds from specialist compounding firms.

Dipping is the process used in making rubber gloves, condoms, medical catheters and balloons. The manufacturing stages for dipped goods are: a shaped *former* made out of glass or ceramic material is dipped into a bath of coagulant, then into a bath of prevulcanized latex on a continuous production line. Drying is followed by vulcanization. Production of examination and surgical gloves for the medical profession is frequently integrated with the manufacture of catheters and condoms. These products are generally marketed through specialist companies supplying the medical, hospital and healthcare industries.

Extrusion is employed in the manufacture of latex thread. The latex is extruded through glass capillary nozzles and the fine threads are drawn through long tunnels where drying and vulcanization take place. The latex thread industry has its own specialist technology and production techniques, and sells its output to textile firms that produce elasticated threads for the garment and clothing industries.

Foam rubber goods are prepared by the aeration of compounded latex followed by the addition of a gelling agent that sets the rubber without breaking down the foam structure. The foamed rubber is shaped in a mould where vulcanization takes place. The major use of foam rubber is in mattresses, pillows and upholstery.

Conclusion

The discussion in Chapter 2 has shown that the consumption of natural and synthetic rubbers is driven by strong demand by manufacturing industry, in particular, the automobile production sector. At the same time, advances in polymer material science as well as rubber processing and manufacturing technology have met the demand from industry for a consistent quality raw material with guaranteed technical specifications to make a range of industrial and consumer products. It is suggested that a basic understanding of these factors helps place in context the detailed examination of the Malaysian rubber products manufacturing sector considered from Chapter 5 onwards. The next chapter, however, is an account of the political economy of Malaysia for readers unfamiliar with historical events that have shaped the character of the country.

CHAPTER 3

Political Economy of Malaysia

Historical Background

The state of Malaysia in its current form came into being a mere 50 years ago with the expulsion in 1965 of Singapore from the newly established Federation of Malaysia. Thus Malaysia today, in political terms, comprises the 11 states of Peninsular Malaysia (Malaya, West Malaysia) occupying the southern extremity of the Malay peninsula of the Asian landmass, and two territories situated on the north coast of the island of Borneo (Kalimantan), the states of Sabah and Sarawak (Borneo states, East Malaysia).

The founding of modern day Malaysia is generally accepted by historians (Andaya & Andaya 2001) to date from the establishment of the kingdom of Malacca (Malay: *Melaka*) in the late 14th century. Malacca was the last of the Malay trading states that had dominated the Straits of Malacca since the foundation of Srivijaya in southern Sumatra in the 7th century. Malacca's dominance may have lasted little over a hundred years (1396–1511) but it was the most important entrepôt, trading in locally produced tin and jungle resources, Chinese goods and spices from the eastern isles. These were exchanged for cotton cloth and other goods imported by Indian and Arab traders from the west. The pursuit for control of the spice trade stimulated European intervention in the Malay world: an area that is now the Philippines, the Indonesian archipelago and the Malay Peninsula. Armed conquest as well as the negotiation of treaties with local rulers by the Portuguese and Spanish, then latterly the Dutch and British for the establishment of European trading posts eventually led to the end of Malay political and economic dominance in this part of Southeast Asia.

The Portuguese conquered the Malacca Sultanate in 1511 but after a number of attacks by the Dutch from 1606 onwards, it fell to a Dutch fleet in 1641. The British occupied the port and immediate hinterland during the Napoleonic Wars (1795–1815) but handed the territory back to the Netherlands after the defeat of France until 1824 when it again returned to British control. Rivalry between Great Britain and the Netherlands led to the negotiation of the Anglo-Dutch Treaty of 1824 that divided control of the littoral to either side of the Straits of Malacca. The treaty allowed the British to pursue their interests in the Malay Peninsula whereas the Dutch followed theirs in Sumatra and Java. The first British possession on the peninsula was the island of Penang (Malay: *Pinang*), which was acquired by treaty in 1786 by the East India Company from the Sultanate of Kedah, a Malay princedom under the suzerainty of Siam. Sir Thomas Stamford Raffles of the East India Company founded the new trading post of Singapore (Malay: *Singapura*) in 1819. In 1826, Penang, Malacca and Singapore were grouped together to form the Straits Settlements, which was a dependency of India until it became a Crown Colony under the Colonial Office in London in 1867.

The opening of the Suez Canal in 1869 made the shipping lane from Europe across the Indian Ocean to the Malacca Strait the fastest route to the East so that the ports on the offshore islands of Penang and Singapore expanded greatly in importance as increased tonnage of steamships passed through them on their way to China, Japan and Australia. At the same time, the industrialization of Western Europe and the north eastern United States significantly increased the demand for tin, a commodity found in abundance in the small Malay states on the west coast of the peninsula. The Malay chieftains who controlled land rights through customary tenure opened mines to exploit the tin ore reserves, and encouraged immigration of coolie labour from the southern provinces of China to work the mines. Capital finance for the mines was invested by influential merchants, both Chinese and British who traded in the British administered enclaves of Singapore, Malacca and Penang. The sudden wealth and conflicting interests of the Malay chiefs boiled over into drawn-out disputes over control of tin-bearing land, while large-scale fighting between rival dialect groups broke out among the Chinese mine workers. With investments and exports threatened, the official British government policy of non-intervention in the Malay states came to an end.

In 1874, the British forced the paramount rulers of the states of Perak, Selangor and Negeri Sembilan to accept British Residents who were charged with providing "advice" to the rulers and their ruling councils on all matters except those regarding Malay religion and custom. The appointment of Residents whose political advice had to be followed, in a system of indirect rule, marked the introduction of British administration and governance across the Malay states. In 1896, Selangor, Perak, Negeri Sembilan and Pahang united to form the Federated Malay States with a centralized system of government in the newly established federal capital, Kuala Lumpur. The kingdom of Siam, which claimed sovereignty over the northern sultanates of Kedah, Perlis, Trengganu and Kelantan relinquished control to Britain in 1909. These states together with the southerly sultanate of Johor retained many features of their own administrations and collectively were termed the Unfederated Malay States. By the second decade of the 20th century, the territory known as British Malaya comprised three distinct forms of government: the Straits Settlements ruled by the Colonial Office in London; the Federated Malay States with a central administration based in Kuala Lumpur; and the semi-independent Unfederated Malay States (Andaya & Andaya 2001: Ch. 1–5).

Economy of Colonial Malaya

The establishment of British political and administrative control over the entire Malay Peninsula allowed stable economic conditions for the development and expansion of the tin mining and natural rubber plantation industries (Allen & Donnithorne 1954; Lim 1967; Jomo 1986; Drabble 2000; Andaya & Andaya 2001). The rubber and tin industries grew rapidly so that by the 1920s British Malaya became the largest producer of both commodities in the world.

The Malays had been mining tin for many hundreds of years but when the surface deposits were exhausted at the end of the 19th century most Malays stopped production. Only Chinese and European mining operations had access to the capital resources to finance the machinery and large workforce needed to extract tin ore below the water table (Lim 1967: 37–71).

The abolition of slavery in the West Indies in 1833 provided the stimulus for the successful establishment of sugar cane estates, owned and managed by Europeans on the mainland across from Penang, using indentured labour from south India. In the late 1800s, there was an influx of European planters from Ceylon after disease had wiped out the coffee crop on the island. The coffee planters settled mainly in Perak, Selangor and Negeri Sembilan where they grew coffee with migrant, south Indian workers until the introduction of the *Hevea* rubber tree as a commercial crop. Chinese farmers practised a form of large-scale, shifting cultivation, growing tapioca (also known as cassava or manioc), or pepper inter-cropped with gambier, a crop used in tanning leather. After the land was exhausted it was abandoned to revert to secondary jungle. This changed when rubber cultivation became an attractive proposition and former tapioca and pepper land was planted with rubber trees and sold to Chinese and European speculators, including London-based plantation companies (Jackson 1968; Tate 1996). Despite a land alienation policy that favoured the establishment of large estates (Drabble 1973: 72–6; Jomo 1986: 66; Jomo 1990: 5) significant Malay peasant participation in rubber cultivation began from 1909, four years after rubber first became a plantation crop on a large scale (Jomo 1986: 64). At this time, all producers planted unselected seedlings and used the same basic production techniques for processing raw rubber.

When British-administered Malaya opened up the economy to foreign investment to produce plantation crop and mining exports, native Malays were comfortably off as peasant farmers in the traditional paddy agricultural sector because of abundant land resources. Malays, therefore, were not willing to work as wage labourers in estates and mines. Thus, large inflows of immigrant labour from southern India and the southern provinces of China were encouraged for the expansion of primary exports. This created a plural society in which equity is viewed as a triangular relationship among foreign investors in mines and plantations, Indian and Chinese immigrant workers and their descendants, and indigenous Malay fishermen and peasant farmers, growing subsistence rice and smallholding rubber (Jomo 1986; Lal & Myint 1996: 143). In his classification of the colonial economy, Junid Saham (1980: 7) regarded there to have been three parallel economic systems of: (a) a peasant economy of rice cultivation, fishing and mixed farming; (b) export production of tin and rubber; and (c) a free trade economy in the entrepôt islands of Penang and Singapore.

Lal and Myint (1996: 108) classified British Malaya as a "land abundant country" in a classification system based on the classical economics school of resources and endowments of land, labour and capital. They defined land abundant countries as those for which capital is scarce relative to both labour and land; labour is scarce relative to land; and land is abundant to both capital and labour. Land abundant countries, therefore, are regarded as suitable for the establishment of large-scale plantation agriculture. In a discussion on the dualistic relationship of plantations and mines with the traditional farming economy in developing countries,

Lal and Myint (1996: 191–2) argued that primary commodities such as natural rubber and tin are produced by large-scale economic enterprises in the modern sector using capital-intensive technology. Mining and plantation firms obtain their labour supply from a well developed labour market based on a regular wage system. They have ready access to modern banking and financial institutions for a supply of capital and credit. Whether foreign or domestically owned, they can draw upon an internationally available supply of managerial and financial resources and up-to-date technology. Provided the available natural resources are sufficiently abundant for profitable exploitation, exports from modern plantation and mining industries can expand, more or less independently of the state of organizational development in the rest of the economy.

Drabble (2000: 109–10), in his economic history of Malaysia, noted that the structure of the Malayan economy in the early 20th century exhibited several aspects of the dualistic paradigm of colonial export economies and indigenous peasant agriculture. There was a clear division into the foreign-owned, capital-intensive, estate and mine sector, and the indigenous, labour-intensive, rice farming sector. However, Drabble also argued that the dualistic paradigm, whilst useful as an analytical device, tends to draw distinctions sharper than those observable in Malaya at this time. The factor making a two sector analysis questionable was the presence of immigrant Asian communities who were becoming increasingly domesticated despite their foreign origins. The Chinese and Indians constituted a layer between the unquestionably foreign western interests whose ownership lay outside the country, and the indigenous Malay population. In the rubber industry, for example, there was a continuum from the larger estates over 400 hectares; smaller estates (40–400 hectares); large to medium smallholdings (10–40 hectares); and the "true" smallholdings under 10 hectares. While the larger estate properties were mostly European-owned, the Chinese were predominant in the smaller estates and the large/medium holdings. The true smallholding sector was predominantly owned by Malays whose rubber farms typically were less than two hectares in area. A similar situation obtained in the mines sector with European-owned companies quoted on western stock exchanges operating highly capital-intensive, floating dredges whereas Chinese and some smaller European mines used the less capital-intensive, gravel-pump technology (Lim 1967: 49–51).

A distinctive feature of the colonial period was that the economy was an open economy operating in a laissez faire system. Apart from government expenditure on ports and rail and road transport infrastructure,

Malaya's development was undertaken almost entirely by private enterprise. Colonial economic practice centred on minimal taxation, strict avoidance of deficits and an essentially unprotected market (Jesudason 1989: 48). Allen and Donnithorne (1954: 51) considered that development under British rule cannot be described solely in terms of western investment. They stated that Asian, especially Chinese, entrepreneurial activity made an immense contribution. The 19th-century establishment of European mercantile trading companies known as Agency Houses, in the opinion of Drabble (2000: 48), nevertheless, greatly increased the scale of trade with the outside world. Merchants both Asian and European spread out from the major ports and were the connecting links that transmitted to the hinterland the "pull" factor of demand emanating from the industrialized countries of the West. Agency House firms in Penang, Malacca and Singapore were active in promoting and funding much of the primary production of tin and natural rubber. Drabble concluded that in the case of Malaysia, trade functioned as the prime "engine of growth" in economic development.

The economy of Malaysia before Independence can be described as a classic case of a colonial economy supplying raw materials to the metropolitan country and providing a market for manufactured imports from the United Kingdom. Wheelwright (1965: 90), for example, described the economy as being heavily dependent on rubber and tin commodity exports, with reliance on imports for most of its manufactured goods, and a substantial proportion of food. Most of the import, export and financial agencies were in the hands of firms largely owned and controlled in the United Kingdom, to which trade was primarily oriented. The dependence on outside sources for essential foodstuffs and manufactured products paid for by export revenue from the trade in rubber and tin is remarked on by Lim (1975a: 2) who regarded the colonial system as one of "imposed free trade" (Lim 1992: 97). Lim (1975a) illustrated his statement with an analysis of import/export statistics towards the end of colonial rule when earnings from rubber and tin accounted for 85 per cent of gross export revenue or 40 per cent of GDP whereas imports of consumer goods, including food, made up 66 per cent of total imports.

Japanese Occupation: 1942–45

The military subjugation of Southeast Asia by Japan in the Second World War (1937–45) was in large measure driven by the aim to gain control of a region considered vital to the Japanese economy (Drabble 2000: 149;

Cooper 2001: 8–10). The war began in 1937 with the Japanese invasion of northern China, which then spilled over to Europe when Germany invaded Poland in 1939 while the USA joined the war in 1941 after the Japanese attack on Pearl Harbor. The Japanese army landed in the north eastern state of Kelantan in December 1941 and swiftly overran the peninsula. Malaya was forcibly incorporated into the Greater East Asia Co-Prosperity Sphere after the fall of Singapore to the Japanese in February 1942. From 1942 to the Japanese surrender in 1945, Malaya was ruled under a military administration centred on Singapore (renamed: *Syonan*). In 1943, the northern Malay states of Perlis and Kedah on the west coast, and Kelantan and Trengganu on the east were returned to Siamese control for the duration of the war (Andaya & Andaya 2001: 257–64, Cooper 2001: 119–20).

The Japanese imposed a planned, command economy based on self-sufficiency in food production and industrialization built around the processing of local raw materials. Exports were to consist of what was required for the Japanese war effort. Rubber production in Malaya and elsewhere in Southeast Asia came almost to a standstill during the Occupation. Rubber growing countries had sold their production almost entirely to the USA and Europe before the war but now the markets were cut off. The Allied blockade made it difficult to ship to Japan which, in any case, was a relative small consumer. Rubber export earnings, which pre-war had constituted the bulk of Malayan exports, were slashed to zero while imports of consumer goods from the West ceased. The standards of living of the majority of the population were drastically reduced and thousands of estate workers were thrown out of employment (Chin 1977: 161–73; Tate 1996: 496; Drabble 2000: 149; Cooper 2001: 328–42). When the war was over, Malayan rubber production recovered very rapidly such that by 1947 production exceeded what was formerly produced before the Japanese invasion (Lim 1967: 86).

Pre-Independence Years

From the end of World War II until Independence in 1957, economic development was characterized by the reconstruction of the economy, particularly recovery from war damage caused by the Japanese occupation, and by further expansion of the export commodity sector (Schätzl 1988a: 29). The Korean War (1950–51) gave a boost to the economy through increased demand for the strategic materials, tin and rubber, and GDP increased by over 60 per cent in one year (Lim 1967: 18; Md Ariff 1991:

171). Political instability throughout the Far East, the wars in Indo-China and the self-imposed isolation of Burma, in the opinion of Md Ariff (1991: 171), gave Malaya an advantage by default when the commercial activities of Rangoon and Saigon moved to Kuala Lumpur and Singapore. In keeping with the received economic wisdom of the time, the colonial authorities, soon after the British reoccupation, set up an Economic Secretariat for the whole of Malaya. The Secretariat produced plans to enable the government to guide private enterprise "to directions which are the most desirable for progress in accordance with decided policy" (Tate 1996: 551). However, as Tate (1996: 552) commented, the "decided policy" was primarily dictated by the needs of the Sterling Area and Great Britain's own economy rather than the requirements of Malaya's post-war reconstruction. On the other hand, Drabble (2000: 165) was of the view that the more dirigiste policy towards the economy of the colonial government was confined to the public sector whereas the stance towards the private sector was for the most part laissez faire. Nevertheless, the reaction of the mainly British business interests to the introduction of economic planning was one of vehement opposition to interference in the traditional preserves of free enterprise (Tate 1996: 553).

The International Bank for Reconstruction and Development (World Bank) was commissioned to report on the development of Malaya and Singapore in 1955 and the Bank's recommendations became the blueprint for the economic and social development of the independent Federation of Malaya in 1957 (Wheelwright 1965: 17; Junid Saham 1980: 8; Jomo 1986: 221). The report noted that the 1953 per capita level of national income at US$250 was the highest in the Far East. Drabble (2000: 160) also commented that in 1950 the per capita GDP of US$1,828, in 1985 constant currency, was easily the highest in Asia, including Japan (US$1,208). He regarded Malaya's leading position in this measure to be primarily due to its high level of export earnings relative to the size of the population. The Paris-based, Organisation for Economic Cooperation and Development (OECD) noted that before Independence the free enterprise, colonial economy was prosperous in comparison with other Southeast Asian countries because of the profitable tin mines and rubber plantations (Lim & Pang 1991: 21). The World Bank also highlighted the importance of rubber and tin exports, and noted that their export value represented about 85 per cent of all exports. The main recommendations of the Bank report were an import-substitution policy to develop secondary industries to supply the domestic market, a programme of agricultural diversification to reduce the dependency on natural rubber, and development of the social and physical infrastructure. In reviewing the economy, the IBRD mission (1955: 13–4) stated:

> By Asian standards, the Malayan economy has reached a relatively advanced stage, not only in the level of per capita income but also in structure: it is a more varied and complex economy than is characteristic of most underdeveloped countries. Power, transport, communications and other basic facilities; a considerable foundation of secondary industry has been established; there is a substantial nucleus of skills and enterprise; standards of public administration are high; and institutional patterns and habits of commerce and finance are correspondingly advanced.

Independence and the Formation of Malaysia: 1945–65

The immediate post-war period was a time of rapid political change, which culminated in the establishment in 1948 of the Federation of Malaya, with its capital in Kuala Lumpur, embracing the former Federated and Unfederated Malay States, and the Settlements of Penang and Malacca. Singapore remained outside the federation as a Crown Colony because of its strategic trading and military importance to British interests (Andaya & Andaya 2001: 264–74). Even before the end of the Japanese occupation, the Colonial Office in London had been making plans to merge all the small states and settlements on the peninsula into an independent Malayan Union. There were plans to include Sarawak and North Borneo in the new Union but the territories were finally omitted and both became Crown Colonies in 1946, while the free port of Singapore with its strategic army, naval and air force bases was not included in the colonial administration's proposals. The plan for the merger into the Malayan Union transferred the tenuous *de jure* sovereignty from the Malay sultans to the British government and abolished many of their privileges. The proposed political settlement gave equal rights and citizenship to all residents in the Union, including the substantial Chinese and Indian communities on the basis of local birth or ten years' residence. There was an immediate and strong reaction by the Malay population against the Malayan Union plan and a new, political force, the United Malays National Organization (UMNO) was established to oppose the Union. The strength of Malay opposition was such that the proposals for union were revoked and the Federation of Malaya was created in 1948. The Federation returned sovereignty to the Malay rulers, enforced more restrictive citizenship rules and gave special privileges to members of the Malay race.

Many Chinese, especially those who had carried on a guerrilla war against the Japanese, felt betrayed by the British and joined the Malayan Communist Party to fight for an independent Malaya under Communist rule. At first, the party waged an open struggle mainly through their control of the trade union movement but later took up armed insurrection that resulted in the declaration of a State of Emergency in mid-1948. However, much of the Communist Party's political support waned when the Chinese-based, Malayan Chinese Association (MCA) together with the Malayan Indian Congress (MIC) joined forces with UMNO to negotiate for full Independence. The "Alliance" comprising the three communally based political parties overwhelmingly won the 1955 Federal elections on a platform of Independence for the Federation of Malaya. Independence from British rule was declared on 31 August 1957 (Andaya & Andaya 2001: 269–82).

Singapore with its majority Chinese population became self-governing in 1958 with matters of internal security and defence still maintained by the British who were due to relinquish control in 1963 when it was expected that Singapore would gain full Independence. The Malayan government, which was still fighting Malayan Communist Party insurgents, was concerned that the communists might take control of Singapore after British withdrawal. The Malayans proposed the idea of a federation comprising Malaya, the Borneo colonies of Sarawak and North Borneo, Singapore and the oil-rich Sultanate of Brunei. During negotiations, Brunei declined to join the new political entity. A consensus was reached among the other four parties with the British authorities, and the Federation of Malaysia came into being in 1963. Political differences arose almost immediately between the People's Action Party (PAP) of Singapore and Malaya's UMNO, which championed the interests of the Chinese and Malays respectively. With a growing threat of communal violence and political battles becoming more heated, the Malay-dominated government enforced Singapore's secession from the Federation of Malaysia in 1965 (Fryer 1970: 209; Lee 1998; Andaya & Andaya 2001: 282–93).

Post-Independence Economic Policy

At the time of Independence in 1957, Malaysia had an export-oriented strategy based on the primary production of natural rubber and tin. The country had "a level of administration, technology, physical and commercial infrastructure and a world trading position that was considerably in advance of most other countries in the South-East Asian region" (Md Zainuddin Salleh & Zulkifly Osman 1982: 125). The economic strategy followed in the years after Independence was essentially a continuation of the free market trade and industrial policies of the colonial government. There was, however, extensive intervention by the government to promote rural development, and to provide social and physical infrastructure, particularly in rural areas (World Bank 1993: 134). Bowie (1991: 69) commented that the development efforts of the newly independent state were focused on the rural-agricultural sector whereas the private sector was left to largely determine the course of development in commerce and industry. Snodgrass (1980: vii) described the overall strategy as one that assigned the development of industry, tin mining and much of commodity tree crop agriculture to the private sector, which was overwhelmingly foreign or ethnic Chinese in its ownership and management. The role of government was seen by Snodgrass to be a passive facilitator of industrialization, whereas its role in rural development was that of a much more active promoter because of government policy intended to improve the lot of the predominately rural Malays. He also stated that the continuation of policies of liberal capitalism coupled with monetary and fiscal conservatism was "what powerful foreign and Chinese business interests wanted, and what the World Bank recommended" (Snodgrass 1980: 47). On the other hand, Jomo (1986: 221) considered that the main departure from the economic policy of colonial times was that the state actively intervened to promote industrialization. Nevertheless, as Cho (1990: 54) and Bowie (1991: 69) argued, the state's role was essentially a passive one by providing sites for industrial development, provision of infrastructure and a benign environment for private investment. The World Bank (1993: 134) stated that while the government did protect domestic industries under its import-substitution programme, rates of protection were low and, importantly, the state did not promote individual sectors.

Agricultural Development Strategy

The newly independent government retained most elements of the pre-war economic patterns including a free market economy inside the country, relying on economic incentives to expand export production from the peasant sector, and from the mining and plantation sectors (Myint 1972: 28–9). Higgins (1982: 148) noted that after Independence, Malaya still had vast tracts of tropical rain forest on suitable land that allowed for a continuation of the policies of successive colonial governments of the expansion of large-scale, modern, technologically advanced, plantation agriculture. A World Bank report (Young et al. 1980: 19) commented on the dualism in the agricultural sector in the 1960s characterized by a modern, estate sector with high productivity and a traditional, smallholder sector featuring low productivity and low incomes.

The development strategy adopted by the government was to diversify the agricultural resource base by increasing the rate of replanting old rubber, new planting of natural rubber and oil palm on logged forest land, opening of new land for rice production, and the improvement of irrigation and drainage in existing rice lands (Chong 1982: 185). During this period, there was increased public expenditure to improve physical infrastructure and provision of social services in health and education in rural areas. Landless Malay farm workers and fishermen were settled on massive land development schemes growing oil palm and rubber on projects undertaken by the Federal Land Development Authority. Other government agencies were established to assist existing rubber smallholders to replant their land, and allow rubber and paddy producers to market their crop (Fryer 1970: 223–65; Jomo 1986; Shamsul & Perera 1977; Shamsul & Lee 1988; Abdul Samad Hadi 1994: 52–5; Lal & Myint 1996: 144; Islam & Chowdhury 1997: 233). As noted by the World Bank in its review of agricultural pricing policies (Jenkins & Lai 1989) and in the analysis by Lal and Myint (1996: 197–8), the investment to pay for these development programmes came from a transfer of resources from the modern, export sector of plantations and mines to the traditional, peasant sector of Malay paddy farmers and rubber smallholders.

A number of commentators including the World Bank (Jenkins & Lai 1989) have noted that Malaysia provided a clear exception to the urban bias and anti-agricultural bias of development theories of the 1960s and 1970s by investing in irrigation projects for rice production, and land settlement programmes for rubber and oil palm cultivation. There is general agreement among these observers that Malaysia avoided an urban-based development strategy because of the political power of rural Malay voters in keeping the Alliance coalition in power (Wong 1979: 99; Jenkins & Lai 1989: 1; Cho 1990: 84; Lal & Myint 1996: 334). On the other hand, in a later, more general World Bank report (1993: 134) it was argued that the government was restrained from implementing incentives against agriculture by the economic and political importance of the "plantation sector". However, the Bank did not say whether the term *plantation* includes the Malay and Chinese smallholding and government-owned land scheme sub-sectors, as well as the technologically advanced, foreign-owned, large estates.

Post-Independence Political Settlement

Underlying the policies of free international trade and a favourable investment climate within a stable financial framework was the ethnic settlement bargained among Malays, Chinese and Indians in 1957. The leaders of the three groups came to an agreement that the Chinese and Indians would recognize the primacy of Malay political power and of the special rights and privileges for Malays, in return for full citizenship rights and a voice in government (Jomo 1986; Andaya & Andaya 2001). This bargain also meant that the Chinese would continue to have economic dominance as long as they did not challenge the political dominance of the Malays (Islam & Chowdhury 1997: 232). Islam and Chowdhury concluded that the communal settlement contributed to Malaysia's economic growth in the first decade of Independence in three ways. Most importantly, it ensured that Chinese economic dynamism was not interrupted; it shielded the state from rent seeking during the import-substitution industrialization phase of the 1960s; and it provided stability that made Malaysia attractive to foreign investment.

The policy of extending the Malay race special rights and political privileges created resentment among the Chinese and Indian communities who also felt that their own cultures were being submerged. This dissatisfaction resulted in the non-Malays voting in large numbers for predominately Chinese-based opposition parties in the May 1969 general election. The Alliance of UMNO, MCA and MIC parties won but lost its two-thirds majority and thus its power to alter the constitution. Severe racial riots broke out in Kuala Lumpur and the fighting, mainly between Malays and Chinese, left 196 dead, according to official figures, with non-Malay deaths exceeding those of Malays by six to one. Parliamentary government was suspended for 18 months and replaced by a National Operations Council that ruled by decree (Bowie 1991: 82; Crouch 1994: 15; Drabble 2000: 196; Andaya & Andaya 2001: 297–9; Kua 2007).

A significant outcome of the riots was a change in economic policy. The previous official hands-off approach towards non-Malay commercial and industrial activities was abandoned in favour of extensive state regulation of the economy. There was, therefore, a marked shift from the relatively laissez faire open economy and social system followed since 1957 to a more authoritarian and interventionist role for government, especially in directing the economy. The new strategy, according to Snodgrass (1980: 60), boldly recognized and confronted the problem of Malay economic inferiority and presented remedies, which stressed movement of Malays to different sectors of the economy and to urban areas. The policy to promote Malay participation was implemented by the creation of numerous state-owned corporations to act as proxies for Malay business enterprises in the economy (Bowie 1991: 83; Gomez & Jomo 1997: 32–8; Drabble 2000: 199). Three Malaysian commentators, Jomo (1986: 300), Khoo (2000: 215) and Kua (2007: 3, 123) also considered that one of the aims of the new economic development strategy was to create a Malay capitalist class and a Malay middle class by using public funds and intervention by the state bureaucracy on a massive scale.

New Economic Policy: 1970–90

The package of measures to improve the economic status of the Malays was termed the New Economic Policy (NEP) and was scheduled to last for a 20 year period, from 1970 to 1990 (Jomo 1986: 256–68; Mehmet 1986: 6–9; Drabble 2000: 197–202; Andaya & Andaya 2001: 301–21). The NEP had two main objectives: (a) restructuring of the economy and society in order to eliminate the close identity between race and economic function; and (b) eradication of poverty irrespective of race. These two aims were to be achieved by a restructuring of employment to reflect the ethnic composition of the population and a redistribution of shares in the national corporate wealth. There was to be no arbitrary division of existing assets but, rather a redistribution of the much larger national "cake" that was expected to result from continued economic growth such that no group would be worse off in absolute terms (Drabble 2000: 197).

When the NEP came to its formal conclusion in 1990, it had, in large measure, achieved many of its objectives and its influence on the economy and Malaysian society continues to the present day. In 1970 it was estimated that some 49 per cent of all households were below the poverty line of RM33 per month (Andaya & Andaya 2001: 302). By the beginning of the present century, the respected *Far Eastern Economic Review* reported that the number of households living below the poverty line had dropped to 5 per cent (*FEER*: 9 Oct. 2003). One significant development during the implementation of the NEP has been the expansion of the middle class, which by 1990 made up almost a third of the workforce. The increase in the Malay component was particularly striking. In 1970 when the NEP was introduced the proportion of the Malay workforce employed in middle class occupations was only 13 per cent, while by 1990 the figure had risen to 27 per cent (Andaya & Andaya 2001: 315). Equally impressive was the increase in Malay employment in the manufacturing sector which, according to Andaya and Andaya (2001: 318), rose from 29 per cent in 1970 to 49 per cent by 1990. There were important changes in equity ownership over the same period. The ownership of corporate assets in 1970 compared to the official figures at the end of the NEP is recorded by Drabble (2000: 197, 202) and shown in Table 3.1, although Andaya and Andaya (2001: 313) noted that there were commentators who considered the Malay share underestimated.

Table 3.1 Corporate Wealth

	1970 (in %)	*1990 (in %)*
Malay	2	20
Other Malaysian	35	45
Foreigners	63	25
Unaccounted	–	10

Source: Drabble 2000.

By the conclusion of the NEP programme, there had been an important shift in the contribution of the two largest sectors of the economy made to Gross Domestic Product. In 1970, agriculture contributed some 30 per cent of GDP but only 16 per cent by 1990; by way of contrast the manufacturing sector rose from 14 per cent in 1970 to 30 per cent of GDP in 1990 (Drabble 2000: 188, Table 10.4). In employment terms, agriculture fell from 48 per cent to 23 per cent of the workforce, and there was an increase in employment in the manufacturing sector from 8 per cent to 29 per cent over the same period (Drabble 2000: 189, Table 10.5).

National Development Policy and Asian Currency Crisis

The successor to the NEP that ended in 1990 was the National Development Policy (NDP). The NDP restates many of the NEP's aims, such as promoting balanced development and optimizing growth, maintaining racial harmony, as well as eliminating social and economic inequalities in society. Under the NDP the emphasis is less on the transfer of wealth and more the rapid development of an active Malay business community, while the mention of ethnic quotas and targets are omitted (Milne & Mauzy 1999: 72–3; Andaya & Andaya 2001: 318). The long-term goal of the NDP is for Malaysia to work towards attaining first world status by the year 2020 by raising the national average income to the same level as highly industrialized countries (Andaya & Andaya 2001: 321).

High growth rates in the 1990s resulted in year on year increases in GDP and corresponding socio-economic gains with a rise in standards of living and full employment (Milne & Mauzy 1999: 75; Khoo 2000: 222). In mid-1997, the Asian currency crisis that began with the collapse of the Thai currency and stock market spread to neighbouring countries in Southeast Asia including Malaysia (Mallet 1999: 142–3; Khoo 2000: 213; Stiglitz 2002: 89). Speculation on foreign exchange markets caused the value of the Malaysian currency, the *Ringgit Malaysia* (RM) to almost half in value from April 1997 to January 1998 while the Kuala Lumpur Stock Exchange share index fell by 40 per cent. By 1998 there was a full scale recession with a severe loss of business confidence and the danger of major conglomerates becoming insolvent. Malaysia refused intervention by the International Monetary Fund (IMF) and introduced a package of capital control measures to insulate the economy from externally generated risks caused by further currency speculation. The *ringgit* was fixed at RM3.80 to the US dollar and its convertibility was abolished so that all international trade had to be conducted in foreign currencies. The fixed exchange rate was maintained till 2005 when a managed floating system was implemented by the central bank although the *ringgit* still is not freely convertible on the international money market. The currency control measures went against IMF policies and Malaysia was severely criticized during the crisis by the international financial community (Andaya & Andaya 2001: 321; Stiglitz 2002: 122). However, in the opinion of Stiglitz (2002: 125): "In retrospect, it was clear that Malaysia's capital controls allowed it to recover more quickly and with a far smaller legacy of national debt burdening future growth than countries such as Thailand, which had followed the IMF advice." Moreover, Stiglitz commented that because overseas corporate investors were primarily concerned about economic stability, there was little evidence that the capital control measures discouraged foreign investment but, on the contrary, investment from overseas actually increased.

New Economic Model

The Western financial crisis of 2008 caused by the imminent collapse of the New York and London international banking sector and the prospect of default of sovereign debt in peripheral economies of the Eurozone currency area had global ramifications. Severe austerity measures introduced by the USA and European nations to balance their budgets led to reduced demand for manufactured goods and commodities in Malaysia's main overseas markets. The subsequent downturn in the economy resulted in the government announcing the New Economic Model (NEM) in order to stimulate economic activity and attract private investment.

The NEM is to be achieved through an Economic Transformation Programme (ETP) in 12 national key economic areas, with the objective of Malaysia becoming a developed and globally competitive economy by 2020. The goals of the NEM are:

- Gross national income of between US$15,000 and US$20,000 per capita by 2020;
- All communities to fully benefit from the increasing wealth of the country; and
- To achieve sustainable development by preserving the country's natural resources and environment while maintaining stringent fiscal discipline.

The overall aim of the NEM is for real GDP growth to increase by a minimum of 6.5 per cent each year until 2020 (Malaysia: NEAC 2009).

The policy initiatives announced under the NEM include an economic liberalization programme by allowing foreign capital to hold majority stakes in most manufacturing and financial enterprises, and lowering the minimum quota for Malay ownership in stock market listed companies from 30 per cent to 12.5 per cent. There is also a reduction in government subsidies on essential household items such as sugar, LPG used for cooking, and petrol and diesel motor fuels that affect the wider population.

Conclusion

Government development policies from 1970 onwards have focused on an industrialization strategy emphasizing export-oriented industries and encouragement of foreign investment. Since 2000, growth rates have relied heavily on the manufacturing and service sectors that represent the country's strength after the Asian currency crisis. Manufacturing remained the fastest growing sector throughout the 1990s to the present day so that currently the sector contributes some 33 per cent of Gross National Product. Politico-economic policies introduced under the NEP and continued in the current NEM programme have resulted in a redistribution of economic wealth to the Malay ethnic group coupled with a consolidation of Malay political control. Nevertheless, the standard of living of all Malaysian citizens, regardless of ethnicity, has improved up to the present time and the country is classed by the World Bank as a high middle income country. According to the Ministry of Finance, the purchasing power parity per capita income in 2012 was US$15,676 with more than 60 per cent of households owning luxury consumer goods such as motor cars and television sets.

Since Independence in 1957, there has been an unprecedented change in the economy of Malaysia from a country largely dependent on the export of primary commodities to a newly industrialized economy exporting mainly manufactured goods. Social changes have been equally important with the majority of the population becoming urbanized and obtaining employment in secondary manufacturing and the tertiary service sectors rather than in the primary sector of tin mining, fishing, peasant farming and plantation agriculture.

CHAPTER 4

Development of Manufacturing Industry in Malaysia

Industrialization under British Rule

An examination of the literature on the development of the manufacturing sector in Peninsular Malaysia indicates that there are two schools of thought regarding the influence of colonialism on the industrialization process. One school of economists argues that under colonialism a dual economy developed: an externally oriented, primary producing sector centred on tin mines and rubber plantations; and a sluggish, Malay, subsistence economy largely outside the orbit of direct British control and capitalist relations of production. The conclusion of this model is that the colonial economy supported by the policies of the colonial state severely inhibited Malaysia's industrialization until well after Independence (Overton 1994: 35). The leading advocate of this school is Jomo Kwame Sundaram, (known in the literature as Jomo K.S.) who unequivocally states that industrialization began with Independence in 1957 (Jomo 1993a: 1). However, this argument is confounded, to a large degree, by whether or not Singapore is to be regarded as part of British-administered Malaya prior to 1957. A report prepared by the World Bank that examined the economy of Malaya in 1955 stated that Malaya comprised the Federation of Malaya and the Colony of Singapore. The report then made separate recommendations on economic development for the Federation and Singapore (IBRD 1955). Surprisingly, Jomo (1986: vii–viii) also argued that: "Economically, culturally, and, at times, even administratively, the now independent city-state of Singapore (Singapura) was also very much part and parcel of Malaya, and the term 'Malaya' often refers to the island as well." The first Prime Minister and architect of Singapore's independence, Lee Kuan Yew (1998: 21) in his autobiography commented that Singapore and Malaya had always been governed as one territory by the British, with the Malayan peninsula being regarded as Singapore's hinterland. Indeed, Lee (1998: 515), writing about the formation of Malaysia in 1963, stated that his policy, at the time, was that Singapore should become the industrial hub of the new country.

These arguments suggest that before Singapore's expulsion from Malaysia in 1965, it was difficult to separate clearly the industrial development of the two territories while they were under British rule. Indeed, Brookfield (1994a: 6) regarded any attempt to view the economy of the peninsula in isolation from that of Singapore before the late 1960s as illusory. Nevertheless, from a present day, Malaysian perspective, it is argued that the industrialization of Singapore had an inhibiting influence on the development of the manufacturing sector in Peninsular Malaysia. This Kuala Lumpur-centric viewpoint is put forward by Jomo (1993a: 1) when he wrote that Singapore was the regional centre for the British Empire where most industries for Malaya were concentrated, and when Malaya achieved Independence in 1957 without Singapore "the newly independent hinterland lost much of its modest industrial sector". In extending the argument, Jomo (1993b: 290) contended that the industrialization of Singapore under colonialism did not contribute to, but pre-empted and discouraged the development of manufacturing in the Malayan hinterland. The importance of the manufacturing sector in Singapore within newly established Malaysia was remarked on by Drabble (2000: 186), who noted that the forced withdrawal of Singapore in 1965 removed much industrial capacity from the country.

In addition to the "Singapore factor" referred to above, a number of commentators have provided reasons for the lack of encouragement of industrial development under the laissez faire policies of colonial rule. Jomo (1993a: 1), for example, cited British imperial policy, decided in London, for the colonies to supply raw materials to British industry and provide a market for manufactured imports from the United Kingdom. In Malaya itself, the colonial authorities discouraged local manufacture, according to Jomo and Edwards (1993: 18), because domestic production would reduce the revenue collected by import duty when products entered the country. The influence of the powerful rubber and tin primary industries is also regarded as an important factor in preventing the establishment of a manufacturing industry base. The largely British-owned rubber and tin interests argued that protection of local consumer industries would raise the cost of living for plantation and tin mine workers which in turn would bid up wage rates, thus reducing the profitability of the two most important sectors of the economy (Wheelwright 1965: 97; Jesudason 1989: 48; Edwards 1992: 157; Jomo & Edwards 1993: 18; Jomo 2007a: 6). In addition to the policy of the import of cheap consumer goods in order to keep wages low, Wheelwright (1965: 97) considered that the labour-intensive rubber plantation industry took the attitude that an expansion of manufacturing would offer alternative employment for estate labour, and so increase pressure on plantation wages.

The alternative school to the one that regarded the industrialization process to have been inhibited under the colonial regime argues that the two primary industries of rubber and tin provided a firm industrial foundation for rapid expansion of the manufacturing sector after Independence. In a review of industrialization in colonial Malaya, Rasiah (1995: 45–68) examined the manufacturing linkages that evolved from the two major export industries, and challenged the argument put forward by the "anti-colonial" school. Rasiah argued that foreign investment in tin and rubber production played an important role by providing effective demand for inputs, while, at the same time, encouraging the diffusion of industrial technology that led to considerable manufacturing growth during the colonial period. Rasiah and fellow commentators such as Thoburn (1973/75, 1973/75a, 1977), Brookfield (1994a) and Overton (1994) advanced the viewpoint that the precursor of the modern manufacturing sector emerged under the open economy and laissez faire conditions of colonial government. There are two main thrusts to this argument. The first is that the expanding primary sector stimulated the development of both backward and forward linkages to a nascent manufacturing sector. Rubber and tin production provided a market for manufactured inputs (backward linkages), such as engineering products made by local firms, that had the natural advantage of proximity of location to plantations and mines. Furthermore, the production of natural rubber was a source of raw material for the establishment of downstream industries (forward linkages) that made simple rubber products, such as rubber-soled shoes, and bicycle tyres and inner tubes, using labour-intensive technology in small manufacturing enterprises. The second thrust to the argument is that rubber estates and tin mines were a training ground for skilled workers such as mechanics and clerical staff, as well as a larger unskilled, industrial labour force that learned ways of working which were easily transferable to the expanding manufacturing sector.

In pre-Independence Malaysia, the major industries were processing facilities for milling estate and smallholding raw rubber into export-grade material, and smelting tin ore into tin ingots (IBRD 1955: 11; Wheelwright 1963: 212; Drabble 2000: 136). In an early study on western investment in colonial Malaya, Allen and Donnithorne (1954: 216) traced the establishment of the modern engineering sector to the late 19th century, when a British company opened a shipyard in Singapore that undertook repair work and built small craft for the coastal trade. In the first decade of the 20th century, the company expanded its operations into the peninsula, especially the capital Kuala Lumpur, and Ipoh which was the centre of the Kinta Valley mining industry, to manufacture tin dredges and rubber milling machinery. The manufacture of tin mining and rubber processing machinery required the establishment of iron and steel foundries and associated engineering workshops where this equipment was fabricated (Allen & Donnithorne 1954: 216; Thoburn 1977: 201; Rasiah 1995: 59). With the growing demand from the tin and rubber production industries as well as copra and palm oil processing from coconut and oil palm agriculture, there were substantial levels of technology transfer from Western-owned engineering firms to local Chinese enterprises, through the sub-contracting of manufacture and construction works (Thoburn 1977: 186–206; Rasiah 1995: 67) so that at Independence in 1957, Peninsular Malaysia had a fair number of small-scale, metal engineering firms, owned primarily by ethnic Chinese (Rasiah 1999: 231). In addition to engineering inputs, the requirement for fertilizers by the plantation industry led to the establishment, in the 1930s, of a plant that produced fertilizer mixtures from imported materials (Puthucheary 1960: 56; Tate 1996: 407). Natural rubber production provided a source of raw material for the development of forward linkage activities through the establishment of downstream rubber goods industries. This topic is discussed at length in Chapter 5.

The second plank of the school that considers the rubber and tin primary sectors in promoting linkages with secondary industries revolves around the transfer of technological skills to the workforce. From their establishment, rubber plantation and tin mining companies operated as industrial concerns that required capital, managerial expertise, wage labour, specialization of production and a degree of technical knowledge (Graham & Floering 1984; Goldthorpe 1987; Overton 1994: 36; Lal & Myint 1996: 191). Thoburn (1977: 210) argued that if a new technology provides local people with the ability to act as an industrial workforce, it has both increased factor productivities and generated externalities.

The skills of fitters, electricians and other technically trained workers are useable by almost any manufacturing industry, while the expertise of managers, supervisors and clerical staff are also transferable to industries outside of tin and rubber production. When it comes to the mass of estate labour, Thoburn (1977: 216) considered that rubber tapping skills are of little use outside the plantation industry. On the other hand, Overton (1994: 45) argued that the labourers working on plantations constituted a disciplined, wage earning, rural proletariat well suited to the demands of an expanding, industrial production sector. In a similar vein, Higgins (1982: 148) commented that, in sharp contrast to neighbouring countries, Malaysia at Independence had half of the labour force already in the modern plantation and mining sectors.

Industrialization in Post-Independence Malaysia

The most comprehensive accounts of manufacturing during the transitional period from colonial Malaya to Independence are by the World Bank (IBRD 1955) and Wheelwright (1963, 1965). At the end of colonialism, foreign investment was still largely concentrated in the primary sector, although, as Rasiah (1995: 45) argued, a significant share of the manufacturing industry had developed from the linkages generated from rubber and tin production. The World Bank mission considered that the manufacturing sector of the time was relatively well advanced by Asian standards (IBRD 1955: 11), and classified secondary industries under four categories: handicrafts; processing of primary commodities (tin, rubber, palm oil and coconut); engineering including machinery for tin mines and rubber processing; and other manufacturing. The "other manufacturing" sector included the local manufacture of bricks and concrete building materials, cement, metal cans and glass bottles, soap made as a by-product from copra milling and palm oil production, and the production of rubber goods (IBRD 1955: 304). In 1955, manufacturing and construction accounted for 5.3 per cent of GDP and roughly 10 per cent of the labour force. Most of the firms, approximately 80 to 90 per cent, were in Chinese ownership and the typical firm was small, with less than 20 workers and privately owned (Drabble 2000: 169).

The 1955 report of the World Bank recommended a limited import-substitution policy to build up secondary industries to cater for the domestic market for consumer and manufactured goods. The mission did not consider that Malaya possessed any comparative advantages for large-scale industrialization and cited lack of a complex of basic raw materials for advanced manufacturing, the absence of local sources of cheap power and fuel, and relatively high wages (IBRD 1955: 84; Drabble 2000: 169). The government accepted the Bank's recommendations, and in 1958 enacted the Pioneer Industries Ordinance with the objective of encouraging industrialization. The main features of the act were exemption from corporation tax for five years to manufacturers whose production was new to the country and tariff protection for the new industries. The government created two independent agencies: the Tariff Advisory Board and the Federal Industrial Development Authority (later the Malaysian Industrial Development Authority, or MIDA) to administer applications for pioneer status (Lo 1972: 74–90; Shepherd 1980: 182).

The World Bank's viewpoint (Young 1980: 84) of Malaysia as a location for investment in the manufacturing sector in the period immediately after Independence was positive because of political stability together with minimal interference by government, ample availability of finance, good infrastructure and a plentiful supply of well educated labour. The official attitude to heavy foreign participation in manufacturing industry, at the time, was positive and policy did not discriminate against overseas capital. Foreign investors were generally thought to represent a source of capital, technology and business skills which could be tapped if manufacturing was to grow rapidly (Snodgrass 1980: 210). Over the period 1958–68, foreign firms in virtually all industries relocated to circumvent tariffs, enjoy tax holidays and benefit from entry to the domestic market, so that in the space of ten years foreign-owned companies dominated in the manufacturing sector (Rasiah 1995: 105; UNCTAD 1998: 21).

Drabble (2000: 169) questioned why Chinese entrepreneurs did not play a more prominent role in the pioneer incentive programme in view of their predominance in manufacturing at the time of the country's independence. He suggested that the major reason lay in the type of corporate ownership and organization structure encouraged by the authorities that followed the pattern of established British Agency Houses and multinational companies. The incentives of the Pioneer Industries Ordinance favoured more capital-intensive industries that employed skilled workers (Thoburn 1977: 160; Verbruggen 1987: 365), compared to the traditional, Chinese-owned, labour-intensive, family businesses.

Osman-Rani (1982: 263) concluded favourably that the results of the import-substitution policy for economic growth included some degree of diversification, a move into manufacturing, a reduction of dependence on imported goods and the creation of some employment opportunities. He noted that the import of consumption goods such as food, beverages and consumer durables decreased by almost half, from 47 per cent in 1961 to 27 per cent in 1970. However, Osman-Rani also pointed out that most new industries were based on imported materials or imported intermediate goods so that there was a low ratio of value added and poor linkage effects with the rest of the economy. In a critical analysis, Jomo and Edwards (1993: 19–24) and Jomo (2007a: 6–7) stated that the import-substitution policy encouraged economic rent seeking through tariff protection, the remittance of profits out of the country by foreign firms and high domestic prices of goods manufactured in Malaysia.

Export-oriented Industrialization under the New Economic Policy

In a small country such as Malaysia, the domestic market began to show signs of saturation after a decade of rapid industrialization (Chong 1982: 185; Verbruggen 1987: 365; UNCTAD 1998: 21) and as the 1960s drew to a close it became apparent that this type of policy provided insufficient employment opportunities for a growing labour force. This led to a shift to an export-orientation in industrial strategy, signalled by the replacement of the Pioneer Industries Ordinance with the Investment Incentive Act in 1968. The 1968 Act provided tax incentives based on criteria such as export performance and the degree of local content in manufactured products (Shepherd 1980: 182). The New Economic Policy formulated in 1970 in order to solve the communal problems that presented the primary threat to the future of the nation guided much of the thinking and planning for the economy. Its impact on the politics and economic development of Malaysia has lasted to the present day.

"Malayanizing" the economy had been a goal of the three five-year development plans introduced between 1956 and 1970 but there had been little progress in eliminating the pre-Independence pattern of foreign and local Chinese ownership of the corporate sector. The NEP addressed the problem of maintaining economic growth while ensuring that more resources and more opportunities became available to Malays. The policy had two principal objectives: firstly, a reduction and eventual eradication of poverty irrespective of race; and secondly a restructuring of society so that identification of race with economic function would be reduced and ultimately eliminated. The government assured the non-Malays that restructuring would occur through sustained economic growth, not through redistribution of existing resources, so that "no particular group experiences any loss or feels any sense of deprivation in the process" (Andaya

& Andaya 2001: 303). While the NEP had an economic dimension, in the opinion of Milne and Mauzy (1999: 51), its declared objective was political with the ultimate goal of achieving national unity, whereas Jesudason (1989: 74) regarded the NEP as an important tool for consolidating Malay political power by increasing Malay control of business. The place that industrial policy had in the overall scheme, therefore, was not solely the maximization of economic growth. More fundamentally, it was to modernize and urbanize the Malays under the social and economic restructuring objectives of the New Economic Policy. The objectives of industrialization were to transform the modern sector in two ways: firstly, to change the racial composition of employment; and secondly, to restructure the control and ownership of assets on behalf of the ethnic Malays (Mehmet 1986: 75).

The switch to export-oriented industrialization coupled with the political imperatives of the NEP to give employment opportunities to rural Malays provided a fresh impetus to industrial growth. Two main types of export-oriented industry were encouraged. There was promotion of natural resource-based exports based on the older, well established production of rubber and tin, and the newer, expanding primary commodities of palm oil and tropical timber. More importantly, however, was the development of a labour-intensive, light industrial sector that manufactured textiles, footwear and garments, and assembled electrical and electronics goods from imported components. The establishment of export processing enclaves in the form of Free Trade Zones and Licensed Manufacturing Warehouses led to the rapid uptake of these facilities by transnational corporations. Multinational companies, particularly those assembling electrical and electronic products, were attracted by an environment that allowed 100 per cent foreign ownership, exemption from custom duties and provided generous fiscal incentives for export. There was, in addition, an abundant supply of Malay migrants from the countryside into the urban, industrial sector that kept labour costs low (Osman-Rani 1982: 270; Lim & Pang 1991: 37–9; Jomo & Edwards 1993: 25; World Bank 1993: 135; Tham & Mahani 1999: 56; Jomo 2007a: 9–12).

The discovery and exploitation of substantial offshore reserves of petroleum oil and gas in the mid-1970s and early 1980s coupled with firm commodity prices for rubber, palm oil, tropical timber and tin had a positive impact on government revenues. Windfall profits from petroleum exports were channelled into state-owned enterprises such as the National Trading Corporation (Malay: *Perbadanan Nasional Berhad*, acronym Pernas) and National Equity Corporation (Malay: *Permodalan Nasional*

Berhad or PNB) that were used to purchase controlling interests in foreign companies in the primary sector. The largest plantation and industrial conglomerate in Malaysia, Sime Darby, which was listed on the London, Singapore and Kuala Lumpur stock exchanges, was taken over by Pernas in 1976 through a boardroom coup. In the early 1980s, PNB made successful, hostile takeover bids on the London Stock Exchange in flagship British companies such as the Guthrie plantation group and London Tin. Other plantation companies, including those controlled by Danish and French capital as well as British companies, entered into negotiations with PNB and other quasi-government enterprises to relinquish ownership to Malay interests (Lim 1988: 19–26; Van Helten & Jones 1989: 179–86; Jesudason 1989: 84–97; Tate 1996: 579–98; Gomez & Jomo 1997: 38; Drabble 2000: 199; Andaya & Andaya 2001: 312–3). Commenting on the takeover of British plantation companies in the implementation of the NEP, Barlow (1996: 599) stated that shareholders had little to complain about in the methods chosen or the compensation paid. Non-Malaysian shareholders who wished to dispose of their holdings were able to do so at prices determined by open market activities on the London Stock Exchange at a time of high commodity prices. By handling the plantation sector in this way, Malaysia reinforced its credibility and good reputation in international financial circles. Barlow concluded that it was this international financial credibility that made the country so successful in its bid to attract the vast sums of capital investment required to finance Malaysia's thrust into manufacturing.

In a perceptive analysis, Jesudason (1989) examined the impact on the Malay/Chinese ethnic divide on the industrial strategy followed by the Malay-dominated government in the implementation of the New Economic Policy. He argued that the policy of the state was to subsume economic development under politico-cultural objectives of expansion of the political power of the Malay elite, while realizing the aspirations of the mass of Malays to elevate their economic status. After the formulation of the NEP, the Malay political leaders and bureaucrats wanted Chinese businessmen to open up ownership and employment opportunities to the Malays. Chinese business, as represented by the Malaysian Chinese Association political party and the Chinese Chambers of Commerce, on the whole resisted this policy and chose not to invest in manufacturing. The larger Chinese enterprises expanded domestically in property and hotel development, plantation agriculture and equity investment on the Kuala Lumpur Stock Exchange as well as stock markets overseas. Small to medium Chinese businesses, on the other hand, chose not to invest at all in order to avoid compliance with the provisions of the Industrial Coordination Act, 1975 which regulated employment and investment under NEP objectives (Shepherd 1980: 184–5; Jesudason 1989: Ch. 5; Lim & Pang 1991: 22; Drabble 2000: 199; Andaya & Andaya 2001: 312).

In examining the role of multinational companies, Jesudason (1989: 167) argued that although the stated aim of the NEP was to reduce the influence of foreign-owned businesses, the Malay leadership encouraged foreign investment to help realize their political ambitions. The Malay elite favoured multinational companies because:

- They were an alternative source of investment to the Chinese business sector.
- Foreign companies were concerned solely with business factors and did not enter the political debate on the future of Malaysian society.
- Compared to the traditional, family-owned businesses of the Chinese, multinational firms, operating in a number of overseas countries, found it easier to comply with NEP policies of increased ownership and executive opportunities for Malays.
- Multinationals provided high levels of employment for rural Malays.

The thrust of Malay strategy, according to Jesudason, was to replace the role of foreign investment in the technologically simple sectors such as banking, plantation agriculture and tin mining, and to co-opt multinationals in the technologically more demanding, manufacturing sector. Mohamed Ariff (1991: 124) also argued that Malaysia regarded direct foreign investment as a major conduit for technology transfer because the country needed foreign technology more than investment by overseas capital.

Other commentators, including Snodgrass (1980: 210), Bowie (1991: 132), Jomo (1993a: 11), Lal and Myint (1996: 280), and Gomez and Jomo (1997), had concurred with the argument that Malay economic nationalism has sought to limit the economic dominance of the Chinese by using the financial power of foreign direct investors as a counterweight to entrenched Chinese business interests. In a critical comment on industrial strategy under the NEP, Jomo (1993b: 297) argued that the "ethnic obsessions" of the Malay-dominated government have discouraged viable projects funded by Chinese domestic investment in favour of industrialization under foreign auspices. Lim and Pang (1991: 38), on the other hand, noted that the constraints of the NEP and the government's sensitivity to nationalistic sentiments curtailed the spread and dominance of foreign investment in the domestic-oriented and resource-based industries where foreign ownership was restricted. Similarly, Taylor and Ward (1994: 106) commented that although Malaysia has depended heavily on foreign investment to achieve rapid industrial growth, there have been two countervailing forces affecting such investment: the desire to industrialize and the need to extend local ownership.

Heavy Industries Promotion: 1980–85

Up till 1980, manufacturing, apart from the resource-based industries, had concentrated on processing imported raw materials, food and chemicals, and assembling imported components, such as electronics for export, and motor vehicles for the domestic market. The then Minister of Trade and Industry, Dr Mahathir Mohamed did not consider that this was a pattern conducive to economic progress and industrialization (Milne & Mauzy 1999: 64), and was concerned that the country was too dependent on primary commodities (Jesudason 1989: 118). Under the direction of Mahathir, who became Prime Minister in 1981, Malaysia launched an ambitious, heavy industries policy with the establishment of the Heavy Industries Corporation of Malaysia (HICOM) with the remit to plan, identify, initiate, invest in, and manage such large-scale, manufacturing projects. The intent was to form joint ventures with foreign corporations in the creation of heavy industry, and to promote technology transfer and the training of a skilled labour force (Andaya & Andaya 2001: 313).

The heavy industries strategy, known as the "Look East" policy, was inspired by the examples of Japan and South Korea, and was designed to give state agencies the leading role in the operation of large-scale, capital-intensive, import-substituting industries. The new industries would make use of cheap energy from offshore oil and natural gas fields and hydro-electricity power to produce intermediate industrial goods and consumer durables for the domestic market. Investment came from a combination of public funds through HICOM, petroleum profits from the National Petroleum Corporation (Petronas), and private foreign investment capital, especially from South Korea and Japan. East Asian technology was sourced through multinational firms, such as Mitsubishi, in joint ventures to establish a national automobile project to produce the Proton motor car, an iron and steel plant, cement factories, and motor cycle engine plants. The policy had pronounced import-substitution characteristics with cheap government credit and other subsidies, together with heavy tariff protection. The intermediate objectives were technology transfer and creation of linkages with other sectors of the economy, while the long-term goal was that of export competitiveness. Most of the HICOM industries suffered heavy losses because of high production costs, stiff international competition from other multinationals, gluts in world markets and massive debts. Then in 1985 there was a collapse in commodity prices that brought about a severe recession in Malaysia (Mohamed Ariff & Semudran 1990: 42; Mohamed Ariff 1991: 11; Bowie 1991: 111; Jomo & Edwards 1993: 28–9; Crouch 1994: 22; Rasiah 1995: 107; UNCTAD 1996: 5; Godement 1997: 6; Tham & Mahani 1999: 57; Drabble 2000: 200; Khoo 2000: 216; Felkner 2001: 134; *Far Eastern Economic Review* 2003; Jomo 2007: 12–4).

First Industrial Master Plan: 1986–95

The failure of the heavy industrialization policy to generate high economic performance led the government to commission the United Nations Industrial Development Organization (UNIDO) to recommend a detailed sector-by-sector strategy to guide industrial development (Anuwar 1992: 1; Lim 1992: 97; Giroud 2003: 109). The Industrial Master Plan (IMP) was prepared jointly by personnel from UNIDO and the Malaysian Industrial Development Authority and extended over a ten year period from 1986 to 1995 (UNIDO 1991: 53). The 1986–95 plan focused on changing industrial planning policy from a largely laissez faire approach to a target-oriented one within an open, free enterprise economic system (Anuwar 1992: 32). The plan was highly confidential, subject to the Official Secrets Act, and remains outside the public domain (Anuwar 1992: 34). Jomo (1990: 134), however, stated that 15 of the 22 volumes were released, although he did not say whether availability of the report was restricted to certain favoured academics such as himself, or not. The sensitivity of the contents of the report, however, is in no doubt because it contains "a remarkable combination of sober—even critical—analysis of Malaysia's industrial heritage and current problems" (Jomo 1990: 135). In the opinion of Taylor and Ward (1994: 114), Malaysia had great difficulty in reconciling its industrial development objectives with its NEP goals relating to the restructuring of ethnic employment and equity ownership. In support of this argument, they quoted the following statement on page 17 of the IMP report: "often these objectives mutually pose constraints to each other and reduce the number of policy choices available to the Government" (UNIDO 1985).

Structural weaknesses identified by the IMP in past industrialization policies include lack of local industrial technology capacity leading to over-dependence on foreign expertise; a shortage of experienced engineers and technicians coupled with lack of technical training programmes; excessive protection of domestic industry leading to inefficient working practices and a decline in motivation to upgrade technology and management systems; biases to large firms and capital-intensive industries; weak inter-industry linkages; over-concentration in electronics and textiles in Free Trade Zones; and constraints imposed by NEP restructuring efforts (Cho 1990: 206; Jomo 1990: 140; Anuwar 1992: 9–11).

The ten-year plan for industrialization as stated in UNIDO's mid-term review of the IMP (UNIDO 1991: 51–2) focused on accelerating growth and enhancing the development of selected industries more attuned to Malaysia's resource availability and comparative cost advantage, and moving from an inward-looking, domestic-oriented manufacturing base to one which is export-oriented. The basic strategies recommended in the IMP report are:

- Accelerating outward-oriented industrialization, with a focus on large-scale expansion of manufactured exports.
- Developing resource-based industries for export.
- Diversifying and upgrading non-resource-based industries for export.
- Selectively promoting strategic heavy industries to support export industries.
- Modernization and rationalization of the industrial structure.
- Developing technology and manpower.
- Taking into consideration existing infrastructure when planning the physical location for new industries.

The IMP is a study of 12 industries (seven resource-based, including rubber products, and five non-resource-based) and provides detailed policy packages designed to guide structural change in each sub-sector. The conceptual framework of development is to identify priority products in each manufacturing sub-sector, for example, palm oil products in the resource-based sector, and electronics and electrical equipment in the non-resource-based sector. The major thrust of the strategy for each group of industries was characterized by an outward-oriented industrialization approach. The plan argued for a free trade regime, keeping infant industry protection to a minimum while advocating incentives for import-substitution and the elimination of discrimination against exports (UNIDO 1991: 53). According to Felker (2001: 135), recommendations covered the following areas:

- lists of specific products to be promoted;
- goals for industry structure;
- targeted investment incentives;
- sector-specific institutions for technological and human skills development; and
- strengthening sectoral business associations.

The government implemented the IMP recommendations immediately (Jomo 1990: 141), and introduced the Promotion of Investment Act, 1986 to provide liberal investment incentives to foreign investors. The act relaxed restrictions on foreign equity to permit 100 per cent foreign ownership in projects exporting more than 80 per cent of output, and majority ownership by overseas investors was allowed for firms exporting at least half of their production or supplying inputs to an export company located in a Free Trade Zone (Anuwar 1992: 35; Drabble 2000: 202; Felker 2001: 135; Giroud 2003: 110).

Second Industrial Master Plan: 1995–2005

The success of the First Industrial Master Plan in reviving export-led growth in a period of buoyant global economic activity had by the early 1990s generated new pressures for structural change. The boom had already outstripped the IMP's targets for inward investment, industrial output and export growth. On the other hand, Felker (2001: 145) regarded the IMP to have done little to improve structural weaknesses in the manufacturing sector. There still remained weak local technical capabilities and a shallow industrial infrastructure with few intra-, or inter-industry linkages.

The Second Industrial Master Plan (IMP2) was launched in 1995 for the period 1996–2005. The second IMP proposed no major reversal in industrial policy and recommended a continuation of the promotion of investment, strengthening of industrial linkages, export orientation and improved human resource training programmes. The plan addressed weaknesses in industrial structure, including technological dependency, lack of indigenous technological capability and poor linkages. The structural paradigm recommended is the Porter-model, cluster-based approach whereby strategies focus on the development of clusters of industries with groups of firms acting as dynamic cores of high value-added industries (Tham & Mahani 1999: 58–9; Giroud 2003: 111).

There are three industry clusters:

- *Resource-based*: these are naturally evolving clusters, highly dependent on local factors including inputs, ownership and technology. Resource-based industries include the manufacture of products from rubber, palm oil, wood and petroleum.
- *Policy driven*: these are industries created by government policy interventions that are almost totally dependent on foreign technology. The automobile and automotive components, and steel and machinery industries are examples of policy driven industries.
- *Internationally linked*: export-oriented, multinational corporations situated in Free Trade Zones producing electronic components, electrical appliances and textiles, are examples of this cluster.

Third Industrial Master Plan: 2006–2020

The Third Industrial Master Plan (IMP3) is basically a continuation of the industrial development strategies laid out in the first two plans with an emphasis on export-led growth, encouragement of foreign capital investment, promotion of high technology industries and human resource development through industrial training and scientific education (Malaysia: MITI 2010). In the manufacturing sector, 12 industries are selected for further development and investment. Six industrial groups are resource-based and six are non-resource-based:

Resource-based industries:

- petrochemical;
- pharmaceuticals;
- wood-based;
- rubber-based;
- oil palm-based; and
- food processing

Non-resource-based industries:

- electrical and electronics;
- medical devices;
- textiles and garments;
- machinery and equipment;
- metals; and
- transport equipment

In addition, eight business areas in the tertiary service sector are targeted for promotion. Tertiary service businesses in the IMP3 scheme include education and training, healthcare facilities and the tourism industry.

Specific strategies identified to enhance Malaysia's position as a major trading nation on the international stage include:

- To encourage multinational corporations to establish and expand operations in Malaysia.
- To encourage investment in high-value products through advanced technology and capital-intensive activities.
- To facilitate the development of industrial clusters in the 12 selected manufacturing industries.
- To develop Malaysia as a regional hub for *halal* foods, biotechnology and automotive components.

The long-term objective of IMP3 is to achieve global competitiveness in the manufacturing and service sectors by increasing export trade and improving productivity. Targets are set for manufacturing industries to grow at 5.6 per cent and the service sector at 7.5 per cent annually over the 15-year period of the plan.

Conclusion

The chapter has considered, in some detail, the two opposing viewpoints on the influence of British colonial rule on the development of an industrial economy in Malaysia. One school of thought argues that the policies of the colonial government retarded the industrialization process under the influence of the dominant rubber plantation and tin mining industries. The contrary point of view, and the one that this study strongly supports, is that the rubber and tin sectors in the 1920s and 30s provided a ready market for the development of a local engineering industry making rubber processing and tin dredging machinery. The production of *Hevea* rubber also supplied the raw material for a small-scale rubber manufacturing sector based on labour-intensive technology. These two points are considered in further detail in the next chapter that describes the genesis of the rubber manufacturing industry. Importantly, rubber estates and tin mines employed managerial and clerical staff, skilled workers such as mechanics and electricians, and large numbers of unskilled, manual workers. All grades of workers were employable in other industries as and when they were established because they were familiar with the requirements of routine factory operations, compared to coastal fishermen and peasant farmers fresh from the countryside. This is exactly what has happened over the past four decades as the mass of plantation workers moved to live in small towns to take up employment in new industries located in purpose-built industrial sites. A consequence of the outflow of experienced estate labour is that the tree crop plantation sector these days relies almost entirely on a foreign labour force (estimated at 80 per cent of the total workforce) recruited from Bangladesh and neighbouring ASEAN countries for manual work in the field.

Chapter 1 has introduced the concept of the Porter model of industrial development as it applies to newly industrializing countries. The Porter hypothesis argues that the main role of central government in developing economies is to provide a basic industrial infrastructure and to implement policies to encourage the establishment of groups of firms making similar products. Porter's model was adopted by the Malaysian government as the main framework for industrial development in the second and third Industrial Master Plans with both plans focusing on the development of clusters of firms producing high value goods in selected industries. However, the establishment of Free Trade Zones for the assembly of electronic and electrical equipment in the 1970s predates the Porter model formulated in 1990. In the Free Trade Zones, groups of companies all producing similar products are concentrated in small geographical areas surrounded by security fences and guarded by customs posts. A striking example of this type of development may be seen in the concentration of electronic firms in the Bayan Lepas industrial zone located in the south of Penang Island where international corporations such as Siemens and Hewlett Packard have their assembly plants cheek by jowl with Taiwanese and local companies producing similar electronic components for export.

When the NEP was adopted the role of government in formulating industrial policy changed from a laissez faire system to one based on active state intervention in deciding priorities in investment decisions. In the view of many commentators, the NEP intervention had an overriding political objective to change the racial composition of towns and the urban working class from predominantly Chinese to Malay while restructuring financial control of the industrial sector in favour of the Malay ethnic group. Critics of the NEP argued that the focus on inter-ethnic redistribution led to economic rent seeking by the politically well connected and the development of "crony" capitalism through the emergence of a new *comprador* class of Chinese and Indian entrepreneurs. Direct government intervention was markedly pronounced between 1980 and 1985 with the launch of the heavy industries programme under the "Look East" policy. The strategy was politically driven with scant regard to the professional opinion of economists whose views were ignored by political policy-makers. The failures of the heavy industrialization programme to achieve international competitiveness and generate strong economic performance forced the political elite to change direction. The First Industrial Master Plan was still interventionist in character with the leading role to be taken by government through industrial strategy but the selection of strategic industries for promotion was technically driven compared to the overt political decisions of the heavy industry programme. The emphasis of industrial policy for the manufacturing sector to achieve long-term global competitiveness in an export-oriented business environment was continued in the second plan that ended in 2005 and continues in the current industrial plan. In conclusion, it may be argued that the actions of the Malaysian authorities since 1970 have been strongly interventionist, with government policies deciding the major features of the industrialization programme while relying on private capital to invest in and develop selected industrial sub-sectors.

The chapter has traced the influence of the main political and economic factors on the development of Malaysia since Independence in 1957, to become a country with the status of a Newly Industrialized Economy. It has been demonstrated that industrial development, from colonial times onwards, has been undertaken almost entirely by private enterprise, both local and foreign, working within a congenial political and administrative framework. The success of Malaysia in creating a favourable infrastructure for industrial development since Independence is amply demonstrated by the fact that the country today is among the world's leading exporters of manufactured goods such as semiconductors, room air-conditioners, rubber gloves, and audio-visual equipment. There are over 200 industrial estates and 14 Free Trade Zones where firms export more than 80 per cent of their production even though they import most of their raw materials. In addition, there are a number of specialist, industrial parks equipped with modern facilities for manufacturing activities and R&D companies using advanced technology. The considerable efforts to attract foreign investors have borne fruit so that currently well over 1,000 international manufacturing companies from countries across the world have operations in Malaysia.

CHAPTER 5

Rubber Manufacturing in Peninsular Malaysia

Introduction

The focus of Chapter 5 is to examine developments in the rubber manufacturing industries in Malaysia since *Hevea* rubber became the most important commodity crop in the early 20th century. Initially the scene is set by tracing the growth of the rubber planting industry and development of the associated rubber processing machinery industry. The overview of the rubber products manufacturing industry is divided into three periods that reflect the main macro-economic policies and specific industrialization strategies followed from British rule to the present day. The first period is from 1920 to 1970 when laissez faire policies were dominant within an open economic system even though, in the 1960s, some protectionist measures, such as tariffs on imports, were implemented to assist in the establishment of a domestic industrial base. The second phase covers the period 1970 to 1985 when there was greater government intervention in promoting industrialization under the New Economic Policy to realign control of economic activities in favour of the politically dominant Malay ethnic group. The third period in the review is from 1985 to 2005 when the first and second ten-year Industrial Master Plans were implemented and the manufacture of rubber products was identified as a priority industry for expansion within the resource-based sector of the economy. A more detailed assessment of the contribution that the rubber sector has made to the economy over the period 2005 to 2012 is then described. The chapter ends with a discussion on industrial structure in the rubber manufacturing sector and the linkages the sector has with other parts of the wider economy.

Natural Rubber Production

Hevea trees were introduced into the Malay Peninsula in 1877 and rubber swiftly became the leading commercial crop. Individual Chinese and European planters were the pioneers in cultivating rubber but small proprietary estates were soon overtaken in area by larger properties owned by joint stock companies floated on the London Stock Exchange. Malay and Chinese smallholders also enthusiastically planted their land with the new crop despite official disapproval by the colonial authorities.

The rate of expansion of *Hevea* cultivation in the early decades of the 20th century was so rapid that Allen and Donnithorne (1954: 106) described the development of the rubber planting industry as "one of the greatest achievements of Western colonial enterprise". The large area planted with rubber trees, the vast sums invested in rubber production by plantation companies and smallholders, and the huge export revenues produced during colonial times and the immediate years after Independence made rubber one of the twin pillars of the economy. In 1955, the World Bank economic mission, for example, reported that rubber trees occupied about 65 per cent of the entire cultivated area of Malaya, and rubber production contributed approximately 60 per cent of export earnings (IBRD 1955: 10). From the early 1900s until 1991, Malaysia was the world's largest producer of natural rubber, before losing its premier position to Thailand and Indonesia (Barlow et al. 1994: 61). The output of rubber from Malaysia over the past 100 years is shown in Table 5.1.

Table 5.1 Production of Natural Rubber: 1910–2012

Year	*000 metric tons*
1910	6
1920	183
1930	467
1940	603
1950	761
1960	764
1970	1,269
1980	1,530
1990	1,292
2000	928
2010	939
2012	923

Source: 1910–90: Barlow et al. 1994: Table 2.3; 2000, 2010, 2012: MRB 2013.

The current area planted to rubber is 1.04 million hectares with smallholdings accounting for 94 per cent of the acreage. The one million hectares under rubber is in contrast to the five million hectares of oil palm, now the most important plantation crop. Rubber production in 2012 was 923,000 metric tons which accounted for 9 per cent of total world production of natural rubber. Currently, Malaysia is the world's fourth largest producer of natural rubber after Thailand, Indonesia and Vietnam. Exports of raw rubber totalled 771,000 metric tons valued at RM7.86 million that contributed, in contrast to the 1950s, a mere 1.12 per cent of Malaysia's export trade in 2012.

Exports and Imports of Natural Rubber

The main market for Malaysian rubber is China that takes over 40 per cent of exports, followed by the EU with an offtake of approximately 25 per cent of overseas sales. Other important export markets are the USA, South Korea and Iran.

Table 5.2 shows Malaysia's exports of natural rubber by grade from 2000. In 2012 rubber exports totalled 771,000 metric tons, of which SMR grades accounted for 729,000 metric tons or just over 95 per cent of total exports. SMR20 derived from field coagulum and used in tyre manufacture is the most important grade by volume and accounted for approximately half of SMR production. The tonnage of exports of latex concentrate, RSS and other rubbers, such as pale crepe, is insignificant in comparison with exports of SMR granulated rubber. In 1968, shortly after the introduction of the SMR scheme, SMR grades accounted for only 7 per cent by volume whereas RSS exports were 61 per cent; crepe rubber was 17 per cent; latex concentrate 14 per cent; and other rubbers 1 per cent of total rubber exports (Thoburn 1977: 132, Table 6.1).

Table 5.2 Exports of Natural Rubber by Type: 2000–12

	000 metric tons				
Year	*SMR*	*Latex*	*RSS*	*Other*	*Total*
2000	854.1	91.8	9.9	22.1	977.9
2005	1,056.0	56.0	8.0	8.0	1,127.9
2010	838.4	47.5	11.0	3.9	900.8
2012	728.5	31.3	7.7	3.7	771.2

Source: MRB 2013.

Another difference between the production of rubber in the 1960s and the early 2000s was that until the mid-1980s Malaysia basically exported all rubber produced in the country. The current situation is that a decline in rubber production by estates and smallholdings has led to massive over-capacity in processing facilities. At the same time there is now a substantial local demand for rubber in the rubber manufacturing sector. This has forced SMR processors and some rubber product manufacturers to source part of their raw material from neighbouring ASEAN countries and even further afield. Growing domestic demand has resulted in an increase in rubber imports from only 31,500 metric tons in 1985 to 548,000 tons in 2000 and 872,000 tons in 2012. In 2000, the major import was latex concentrate from Thailand that is used in the latex dipped goods sector, whereas a number of producers of general rubber goods imported small quantities of TSR grades. Rubber processors imported mainly raw coagulum together with low grade sheet rubbers and some TSR for milling and upgrading into tyre grade SMR20 material (Abdul Hamid Sawal 2001: 109). However, more recent figures show that the volume of dry rubber imports has become of major importance, reflecting increased demand from processing companies that remill low quality imports into high quality SMR rubber, and increased offtake of imported TSR rubber by the expanding manufacturing sector. Table 5.3 provides details of imports into Malaysia by type between 2000 and 2012.

Table 5.3 Imports of Natural Rubber by Type: 2000–12

	Metric tons		
Year	*Dry rubber*	*Latex*	*Total*
2000	164,241	383,993	548,234
2005	169,137	292,720	461,857
2010	344,446	361,804	706,250
2012	549,333	322,202	871,535

Source: MRB 2013.

Table 5.4 shows the volume and source of rubber imports from 2000 onwards and demonstrates the importance of ASEAN countries as the source of Malaysia's imports, with Thailand supplying over half of total imports.

Manufacture of Processing Machinery

In Malaysia technically specified rubber (TSR) is produced under the Standard Malaysian Rubber (SMR) scheme introduced in 1965. Before the

Table 5.4 Imports of Natural Rubber by Country: 2000–12

	Metric tons		
Country	*2000*	*2005*	*2012*
Thailand	336,260	435,940	498,116
Indonesia	48,072	52,561	10,503
Philippines	32,125	39,468	51,387
Burma	22,514	29,889	19,115
Vietnam	6,730	68,457	167,813
Cambodia	2,672	5,409	23,177
Others	13,484	74,526	101,424
Total	461,857	706,250	871,535

Source: MRB 2013.

establishment of the SMR specifications almost all solid grades were made into Ribbed Smoked Sheet (RSS) and crepe rubbers.

The installation of rubber milling machines on estates for sheet and crepe production began immediately after the establishment of large-scale *Hevea* cultivation (Rasiah 1995: 59). The early equipment consisted of cast iron rollers powered by imported engines (Thoburn 1973/75a: 39; Thoburn 1977: 198) mainly manufactured by small foundries that also produced gravel-pump mining equipment for Chinese-owned tin mines (Thoburn 1973/75a: 36; Thoburn 1977: 193). The businesses involved in this type of light engineering were owned by ethnic Chinese although there were a few larger firms both European and Chinese-owned that manufactured rubber milling equipment but whose main interest was in tin dredge construction (Allen & Donnithorne 1954: 261; Thoburn 1977: 193). In addition, Thoburn (1973/75: 18) noted that minor items of rubber estate equipment such as tapping knives and containers to carry liquid latex in the field were made locally by small Chinese-owned firms. Thus, by the late 1950s there was a well established engineering industry located in and around Kuala Lumpur and Ipoh that made machinery and equipment for the production of RSS and crepe rubbers as well as for the tin mining industry (IBRD 1955: 304; Wheelwright 1963: 212; Wheelwright 1965: 17; Rasiah 1995: 231).

The introduction of new methods of processing raw rubber in capital-intensive, central factories under the Standard Malaysian Rubber programme in 1965 led to the phasing out of small-scale processing on smallholdings and smaller estates that previously produced sheet rubber. It also resulted in major changes in factory design and machinery construction.

Production of rubber processing equipment such as sheeting and creping batteries moved from numerous small workshops into the hands of a limited number of larger, more specialist firms. In the late 1960s and 1970s, a two-tier system developed whereby SMR factories were designed by consultant engineering firms, or the engineering departments of European plantation companies, who subcontracted the actual manufacture of machinery and construction of factory premises to Chinese-owned engineering companies (Thoburn 1973/75a: 36–7; Thoburn 1977: 193, 201–3). Junid Saham (1980: 114) noted that in a parallel development in 1966, the British plantation group Guthrie developed and manufactured the machinery for the *Dynat* method of crumb rubber production as part of the newly introduced SMR scheme. During field work carried out in 1970, Thoburn (1973/75a: 33, Table III) examined the ratio of local to total capital expenditure for a ten ton per day capacity SMR processing factory. The local content was 64 per cent of expenditure with items such as building work, bulking and coagulating tanks, size reduction equipment, driers and baling presses accounting for between 80 to 100 per cent of total cost. As part of the same research study, Thoburn quoted the example of three long established local firms of rubber machinery manufacturers who were producing equipment for SMR processing that accounted for 70 per cent of their turnover. When Indonesia and Thailand introduced their own Technically Specified Rubber quality control schemes, Malaysian engineering firms were well placed to win contracts for the construction of rubber factories including the manufacture and installation of processing machinery (Rasiah 1995: 59).

Currently there are 12 entries for engineering firms that manufacture TSR rubber processing machinery in the trade directories published by the Malaysian Rubber Board (MRB 2006) and on the website of the Malaysia External Trade Development Corporation (METDC 2013). The websites of a number of companies show that all are involved in the design, fabrication, installation and commissioning of TSR plants in rubber producing countries in Asia, Africa and South America. Indeed, the author can confirm from personal experience that "Made in Malaysia" rubber processing factories are to be found in all the countries that he has visited over 35 years of professional life in the natural rubber industry.

Rubber Goods Manufacture: 1920–70

The production of natural rubber from the *Hevea* tree provided a source of raw material for the fabrication of rubber goods and allowed the development of the rubber products manufacturing industry in Malaya. Thoburn (1973/75: 20; 1977: 160) commented that local manufacture of rubber goods was carried out long before the Second World War. In 1926, the British Agency House, Harrisons & Crosfield who were secretaries and agents for a large number of rubber plantation companies invested capital in the Linatex manufacturing company. Linatex, originally called the Wilkinson Process Rubber Company (after the inventor of a method of low temperature vulcanization of liquid rubber) produced abrasion-resistant rubber products used in the local tin mining industry and mining and mineral extraction industries overseas. By the 1930s, Linatex was exporting rubber products to mining companies in North and South America, Europe, Australia and countries in southern Africa (Pugh 1990: 101–4, 120–1, 144, 241–2).

Drabble (2000: 136) reported that in 1923 a Chinese businessman, Tan Kah Kee, one of the richest men in Malaya with interests in rubber, pineapple, rice, sugar and shipping (Coates 1987: 225), established a factory in Singapore to make rubber-soled canvas shoes that exported most of its production to Southeast Asia and China. The first record of the manufacture of rubber footwear in Malaya proper was in 1937 when the Czech company, Bata, that had production facilities in India, began production in Klang when it brought staff from its Indian operations to train local workers (Allen & Donnithorne 1954: 216; Rasiah 1995: 60). Allen and Donnithorne noted that two Chinese firms recruited skilled labour from Bata and opened shoe factories in Klang about the same time. There are two firms, Shum Yip Leong established in 1921 and Fung Keong founded in 1939, located in Klang where the Bata factory was situated. It is a reasonable assumption that these two Chinese-owned businesses are those mentioned by Allen and Donnithorne especially since one company, Shum Yip Leong still makes rubber boots even though its main business today is the manufacture of industrial products. Indeed, the probability is that the Shum Yip Leong Rubber Works was making rubber shoes before the Bata factory was built in view of the fact that it was in business well before Bata came to Malaya. The other firm, Fung Keong Rubber Manufactory currently restricts its business activities in Klang to the manufacture of tyres and tubes for motorcycles and bicycles, inner tubes for commercial vehicles and solid industrial tyres. There are five other firms founded in the colonial era that are still in business today. Nam Bee was founded in 1938 and is a compounder producing master-batch as well as tyre retread materials. The Kinta Rubber Works, established in 1940 in the tin mining centre of Ipoh, is a manufacturer of moulded, extruded and calendered industrial products. Kayel Rubber Products was also established in the 1940s and is a manufacturer of re-treading products for export as well as being a major tyre retreader in Malaysia. In 1956, just before Independence, the firms of Swan Rubber and Sun Yuen Rubber began manufacturing operations. Swan Rubber is a manufacturer of industrial rubber goods and Sun Yuen makes a range of tyre inner tubes for small wheelbarrows through to massive earthmovers as well as tyre retreading materials (MRB 2006; MRPMA 2006).

In 1955, the World Bank (IBRD 1955: 304) reported that the secondary rubber manufacturing sector in the Federation of Malaya included the production of rubber footwear, hoses, belting, and bicycle tyres and inner tubes. The view of the Bank mission was that the production of rubber goods had the advantage of using a local material and Wheelwright (1963: 212), commenting on the World Bank report, stated that the simple rubber products manufactured at the time were labour-intensive operations that involved little capital equipment. In his pioneering study of the ownership of plantation and industrial companies in Malaya, Puthucheary (1960: 131) noted that by the early 1950s there were "about 32" factories making rubber goods of which 31 were Chinese-owned and "some of the Chinese rubber goods factories are quite big and are comparable in size to the large Bata factory in Klang". Rasiah (1995: 60–1) stated that the output of footwear, rubber soles and heels, sheeting, matting, foam rubber goods, tubing and hoses as well as tyres and inner tubes for bicycles grew substantially in the 1950s. The production of footwear, and bicycle tyres and tubes rose sufficiently to enable exports of these products for the first time after the disruption to trade caused by the Second World War. In 1955, Malaya exported 5.3 million pairs of shoes and slippers, and exports of bicycle tyres and tubes grew by 20 per cent annually in the period 1953 to 1955 (Rasiah 1995: 60–1).

The World Bank report of 1955 recommended a strategy for the economic and social development of Malaya that was adopted by the Alliance government after Independence. On the issue of rubber manufacturing, the report did not consider that Malaya could become a major centre for the production of rubber goods because raw rubber is cheap to ship relative to its market price; and it is a small part of the production cost of its most important end use in tyres. The World Bank mission concluded that the most economic location of rubber manufacturing industries is in the major centres of consumption. Nevertheless, the report recommended the local manufacture of rubber goods to supply the home market, especially foam rubber products and the establishment of a single modern tyre factory to produce tyres for motor cars, light vans and trucks (IBRD 1955: 306–7; Wheelwright 1965: 17).

The British rubber manufacturer, Dunlop established a local subsidiary, Dunlop Malayan Industries with a factory in the new town of Petaling Jaya on the outskirts of Kuala Lumpur in 1961. The factory initially manufactured automobile tyres and inner tubes and foam rubber products for domestic sales. It later extended its range to include golf balls and adhesives. The company had pioneer status and benefited from tariff protection on imports of car and commercial vehicle tyres, which was introduced under infant industry protection policies recommended by the IBRD development plan for industrialization (Wheelwright 1965: 92; Thoburn 1977: 160). Other rubber manufacturing companies that started operations in the late 1950s and 1960s include the Sun Rubber Corporation, originally Fock Hee Sun (1957, retread materials); Autoways (1962, retreaded tyres); Yap Hoi Kong (1962, general rubber goods); Universal Cable (1967, insulated cables); Kulitkraf (1968, military boots); Aerofoam and Weifong (1969, foam mattresses and upholstery) (MRB 2006). The economist, Thoburn (1973/75: 19; 1977: 160) working at the University of Malaya reported that by 1970 there were a total of 50 rubber manufacturing companies, including five capital-intensive firms listed on the Kuala Lumpur Stock Exchange, and that the industry employed over 8,000 workers. The total value of the output of rubber products in 1970 was RM119.5 million, divided into domestic sales of RM102.8 million (86 per cent) and exports of RM16.8 million (14 per cent) (Abdul Hamid Sawal 2001: 111, Table 5). The rubber manufacturing sector consumed 22,129 metric tons of raw rubber in 1970 which, according to Thoburn (1977: 160), accounted for 1.3 per cent of Malaysia's rubber production.

Table 5.5 shows the growth in the number of manufacturing firms for the 20-year period, from 1950 to 1970. The number of enterprises involved in the manufacture of rubber products rose from approximately 30 firms before Independence in 1957 to 45 establishments in the years immediately afterwards, then to 50 manufacturing companies in 1970 when the New Economic Policy was introduced. Over the same 20-year period the number of workers employed by the industry expanded from 5,700 to more than 8,000. The censuses taken in 1947 and 1963 provided additional information about the status of the rubber manufacturing sector. The 1947 census indicated that 4 per cent of the labour force was employed in secondary rubber manufacture out of a total working population of just over 143,000 workers, whereas the 1963 census stated that the workforce in rubber products manufacture was 8 per cent of the total employed in the secondary manufacturing sector. The census of 1963 provided figures to show that in addition to the rubber manufacturing industry producing rubber goods, there was also a "backyard" sector involved in the retreading of tyres. This backyard industry consisted of 162 small businesses employing on average only three workers each.

Table 5.5 Rubber Goods Manufacturers: 1950–70

Year	*No. of firms*	*No. of employees*
1950	32	5,700
1963	45	5,565
1970	50	8,375

Source: 1950: IBRD 1955: Table 1, 1947 census; Puthucheary 1960: 131
1963: Lo 1972: Appendix 1, Table 2.4, 1963 census
1970: Thoburn 1973/75, Table V; Thoburn 1977: 160

Rubber Goods Manufacture: 1970–85

After the introduction of the New Economic Policy in 1970, there was a change in direction in government policy on industrialization. The infant industry protection policies advocated by the 1955 World Bank report were replaced by policies designed to encourage export-oriented industries while supporting import-substitution for consumer goods. As tariffs on several pioneer status industries fell, import-substitution gradually lost its significance in terms of output and in the generation of employment (Rasiah 1995: 106). The main emphasis was on the development of a new light industrial sector assembling electrical and electronic goods from imported parts, and the production of textiles and clothing in Free Trade Zones. Of secondary importance was the promotion of resource-based manufacturing using natural rubber, palm oil, tropical timber and tin. One of the first decisions taken under the New Economic Policy in the rubber manufacturing sector was in 1972 for Pernas, the government trading and investment arm, to enter into a joint venture with the American multinational, Goodyear Tire and Rubber Company to manufacture automobile tyres, thus breaking the de facto monopoly of Dunlop Malaysian Industries (Junid Saham 1980: 46).

The number of industrial firms making rubber products increased from approximately 50 in 1970 to 135 by 1985, according to figures published by the Malaysian Rubber Board (Abdul Hamid Sawal 2001: 111, Table 5). This represented a substantial expansion of the sector, with 85 new companies being founded over 15 years, while output value increased five-fold from RM120 million to RM650 million over the same period. The increase in exports of rubber goods during this period was impressive. Exports rose from RM17 million in 1970 to RM215 million in 1985; an increase of almost RM200 million in 15 years. These figures indicate a change in marketing strategy for the rubber industry in line with government policy to promote export production. In 1970 the emphasis was on supplying the local market, with most production being sold within Malaysia and 14 per cent of output being sold into export markets. However, by 1985 overseas sales had increased to one-third of total production. Despite the expansion in rubber manufacturing, Anuwar (1992: 40) noted that rubber goods production grew at a slower rate than the overall manufacturing sector in the decade from 1971 to 1980. He stated that the average annual growth rate of rubber products manufacture increased by 5 per cent, compared to an increase of 11.4 per cent for the total manufacturing sector.

Rubber Goods Manufacture: 1985–2005

In 1985 a severe economic recession struck Malaysia, caused by a combination of a collapse in world commodity prices, failures in the heavy industrialization policy that led to large financial losses in HICOM industries, and a downturn in the electronics business. The bleak economic climate forced the government to reappraise its policies on industrialization and it commissioned the United Nations Industrial Development Organization to prepare the First Industrial Master Plan (IMP1).

The plan advocated an export-oriented industrialization strategy based on the expansion of the production of resource-based manufactured goods in order that the economy could expand in line with government policy objectives. The rubber manufacturing sector was identified as a priority sector in a group of seven resource-based industries where Malaysia was regarded as having a natural comparative advantage. The long-term strategy was to make the manufacture of rubber products one of the leading industries in the resource-based sector, and convert rubber plantation agriculture from a primary commodity exporter into the raw material base for a vertically integrated, export-oriented, manufacturing industry producing intermediate and consumer rubber goods. To this end, IMP1 set a target to increase local industrial consumption of natural rubber from 65,000 metric tons in 1985 to 300,000 metric tons in 1995 (UNIDO 1991: 53; Anuwar 1992: 41; Abdul Samad Hadi 1994: 58). The UNIDO recommendations accorded priority to the inflow of foreign investment on the assumption that overseas capital would provide the greatest impetus for technological upgrading. The role of direct foreign investment was expected to spearhead Malaysia's resource-based industrialization by enhancing the natural comparative advantage of the existing resource base, and producing medium to high value-added manufactured products that could compete successfully on the world market (UNIDO 1991: 62).

The emphasis of the IMP's development strategy was on the expansion of the tyre industry, designating the priority product area within the rubber manufacturing sector, and to raise Malaysia's share in world tyre production to 1.5 per cent by 1995. The strategy also aimed at strengthening the research institutes of the Malaysian Rubber Board to become centres of excellence in applied research for rubber manufacturing in general and tyre manufacture in particular (UNIDO 1991: 53; Abdul Samad Hadi 1994: 58). The IMP's objectives for the rubber products sector are detailed by Anuwar (1992: 47):

- To encourage aggressive export promotion and development of selected key rubber products, particularly tyre and latex dipped goods, by improving the level of competitiveness through adopting cost-reduction measures and increasing productivity and product quality in order to meet IMP export targets.
- To encourage greater foreign investment, especially by multinational corporations, in order to attain greater cooperation with them in R&D activities and to gain access to export markets.
- To place Malaysia at the forefront of R&D in rubber manufacturing as well as in natural rubber production through more financial support to local and overseas institutions.

The second Industrial Master Plan (IMP2) covered the period 1996–2005 and divided its recommendations for the development of the manufacturing sector into three clusters of industries: internationally-linked, policy-driven and resource-based. The manufacture of rubber goods was once again identified as one of the priority areas under the resource-based cluster but, unlike the first plan, IMP2 did not set any specific targets for the rubber manufacturing sector. The plan proposed an overall strategy that development should be technologically based, with the emphasis on complementary linkages with R&D, industrial design, automation of factory operations, distribution and marketing activities. In particular, IMP2 envisaged a greatly strengthened role for government and private research institutes in R&D in rubber technology and the rationalization of marketing efforts in the rubber sector. The second plan recommended diversification of the range of rubber products in order to reduce over-reliance on latex dipped goods, and identified rubber engineering and industrial products as having potential for high added value by the application of advanced technology (Ong 2001).

UNIDO's recommendations for linkages from research through to manufacture and marketing in a vertically integrated, rubber growing and manufacturing industry led to a number of organizational changes in the institutions involved in the regulation of rubber trading and research into natural rubber. The Malaysian Rubber Board (MRB) was established in 1998 from a merger in Malaysia of the Rubber Research and Development Board, the Rubber Research Institute and the Rubber Exchange and Licensing Board, and in the United Kingdom with the Malaysian Rubber Producers' Research Association. The MRB's activities to carry out research and promote downstream industrial activities are undertaken in two operational units, the Rubber Technology Centre at Sungai Buloh, Selangor and the Tun Abdul Razak Research Centre (TARRC) at Brickendonbury, near Hertford in England. The two centres carry out applied research into rubber technology, concentrating on latex and engineering applications, and TARRC undertakes fundamental research in polymer science and the study of elastomers. As part of the MRB's remit to assist in the development of rubber manufacturing industries, both centres carry out specific research on industry problems, and offer consultancy and contract research services for individual manufacturers. The Rubber Technology Centre, in particular, is involved in providing laboratory testing and quality control services for local firms to maintain and improve the quality of rubber products (Ong 2001; TARRC 2007). The Malaysian Rubber Export Promotion Council (MREPC) came into being in 2000. The council's mission is to undertake market promotion of rubber manufactured goods in world markets, with particular support being given to small and medium enterprises. As well as assistance in the promotion of made-in-Malaysia rubber products, MREPC activities include the identification of new export opportunities and facilitating market access in existing export markets (MREPC 2013).

Table 5.6 gives a profile of the rubber goods manufacturing industry over the 20-year period from the start of the first Industrial Master Plan to the end of the second plan in 2005. The table shows that the number of industrial enterprises increased by over 200 firms, from 135 companies in 1985 to 357 by 2005. There was a spectacular increase in the use of natural rubber in rubber manufacturing over this period, from some

50,000 metric tons in 1985 to 352,000 metric tons in 1995, thus exceeding the target set under IMP1, and rising to 483,000 metric tons when IMP2 came to a close. In 2005, natural rubber consumption by manufacturing industry came to 483,000 metric tons, accounting for some 43 per cent of Malaysia's production of 1.12 million metric tons of raw rubber. The value of exports of rubber manufactured goods greatly exceeded those of raw rubber, totalling RM8.03 billion, compared to natural rubber exports of RM5.79 billion in 2005. Export sales of rubber products increased from 33 per cent of the industry's output in 1985 to 80 per cent of production in 2005, in line with the export-oriented strategy recommended in the two Industrial Master Plans.

Table 5.6 Rubber Products Manufacturing Industry: 1985–2005

Year	*No. of companies*	*Employment*	*Rubber consumption*	*Output*	*Exports*
			metric tons	*million (in RM)*	*million (in RM)*
1985	135	18,339	50,291	650.95	215.27
1990	255	36,290	187,592	2,108.44	1,876.67
1995	292	52,885	351,895	4,422.99	3,866.98
2000	319	63,125	419,323	7,089.53	5,685.55
2005	357	63,112	482,889	10,093.80	8,031.00

Source: MRB, 2000–08.

The expansion in the number of enterprises making rubber products between 1985 and 1995 was a result of the sudden upsurge in demand for latex medical examination gloves in the USA and Western Europe following the worsening AIDS crisis. Malaysia, as the world's largest producer of latex concentrate with a well established industrial infrastructure and strong technical support from the Rubber Research Institute, was the most attractive country for foreign multinationals and local companies to invest in medical glove production. As a result the number of firms more than doubled in ten years with the dipped latex goods sub-sector experiencing remarkable growth (Abdul Hamid Sawal 2001: 109). However, as Abdul Hamid Sawal (2001: 110) and Ong (2001: 117) noted, the MRB recognized that the dominance of latex goods production is the greatest weakness of the rubber products industry because of its very narrow product base. It is for this reason that IMP2 recommended the need to broaden the range of products and to promote high value-added industrial rubber manufactures.

Profile of Rubber Sector: 2005–12

At the time of the Western financial crisis in 2008, buoyant commodity prices stimulated by strong demand from China coupled with a downturn in exports of manufactured products led policy-makers to reassess the role of the rubber growing sector as a contributor to economic growth after three or more decades of official neglect in favour of oil palm cultivation. The strategy formulated in the Economic Transformation Programme in the NEM development plan is to increase the contribution the industry makes to Gross National Income from its current level of around RM20 billion to RM53 billion by 2020. The specific target set for the rubber production sector is to achieve an output of two million metric tons per year by increasing the average productivity of the rubber agricultural sector to 2,000 kg/ha/yr and embarking on a new planting programme of 40,000 hectares every year until 2020 (Malaysia: NEAC 2009).

Six new areas of business have been identified for expansion in the rubber manufacturing sector in the Third Industrial Master plan:

- To strengthen Malaysia's position as the world's leading producer and exporter of latex products.
- To diversify the product range in the industrial and general rubber products sub-sector.
- To expand the export market for Malaysian rubber products focussing on South America, Africa and the Middle East.
- To encourage outward investment by Malaysian companies producing low value-added rubber goods to set up subsidiaries and joint ventures in ASEAN rubber producing countries with lower production costs.
- To develop Malaysia as a regional centre for the testing and certification of rubber products made in ASEAN countries.
- To upgrade production technology by factory automation.

The contribution that all sectors of the rubber industry made to Malaysia's export trade over the period 2000 to 2012 is shown in Table 5.7 (using a different data set from Table 5.6).

In 2012, the value of exports of rubber manufactured goods was approximately twice the value of raw rubber exports. When the revenues of export sales of natural rubber, rubber goods, other rubber and *Hevea* timber products are added up, the combined contribution to overseas sales from the entire rubber industry was 5.19 per cent of exports valued at RM36.42 billion.

Table 5.7 Rubber Industry's Contribution to National Exports: 2000–12

Year	*Natural rubber*	*Other rubber*	*Rubber products*	*Heveawood*	*Industry total*
	billion (in RM)	*billion (in RM)*	*billion (in RM)*	*billion (in RM)*	*billion (in RM)*
2000	2.57	0.09	5.51	5.1	13.27
2005	5.97	0.45	8.32	7.25	21.99
2010	9.13	4.24	12.85	7.63	33.85
2012	7.86	6.40	14.53	7.63	36.42

Source: MRB 2013.

The figures in Table 5.7 have a new category of "other rubber" compared to MRB statistics for previous years. In Table 5.7 the other rubber category consists of synthetic rubber, reclaimed rubber, waste rubber, compound rubber and unvulcanized rubber that was an insignificant sub-sector of the industry before 2000. (The other rubber category in historical statistics refers to high specification formulations of modified natural rubber produced by a small number of processors for specialist markets or other grades of raw rubber.) The demand for reclaimed synthetic and natural rubber was led by the local rubber products industry that wished to utilize cheaper sources of raw material than imported synthetic rubber and SMR grades. However, the main increase since 2005 has been in exports of synthetic rubbers from newly established synthetic plants and "China Compound Rubber" because of new market opportunities for sales of compounding material in China. This has resulted in SMR rubber processing factories switching to the manufacture of China Compound Rubber and some existing rubber compounders increasing production for the Chinese market. Chee (2010) reported that 30 firms produce China Compound Rubber with a typical formulation of 97 per cent SMR20 natural rubber with 3 per cent styrene butadiene synthetic rubber (SBR) plus admixtures of carbon black and stearic acid.

In 2012, the MRB records show that there were 349 companies producing rubber manufactured goods that employed more than 60,000 workers. The greatest number of firms at 201 was in the general products sector that includes footwear, industrial and general products, 125 companies produced latex-based products and 23 companies made tyres and inner tubes. The value of production of the sector was RM15.9 million with domestic sales of RM1.4 million.

Table 5.8 shows the value of Malaysia's exports of rubber manufactures by product sector in 2012 which totalled RM14.5 million and accounts for 91 per cent of total production. The most important sector by far is the latex products industry that contributed some 83 per cent to export sales, followed by the general rubber goods sector at 12 per cent, then pneumatic tyres and inner tubes at 5.5 per cent. Malaysian rubber products are exported to more than 180 countries in the world. The four most important markets for rubber goods are the industrialized economies of the USA, European Union and Japan, and newly emerging China that together account for approximately 65 per cent by value of exports.

Table 5.8 Value of Exports by Production Sector: 2012

Tyres/inner tubes	*General products*	*Latex products*	*Total*
'000 (in RM)	*'000 (in RM)*	*'000 (in RM)*	*'000 (in RM)*
578.62	1,387.14	6,159.67	14,527.90

Note: General products include industrial rubber goods, general rubber goods and footwear.
Source: MRB 2013.

Malaysia is the sixth largest consumer of natural rubber in the world after China, Japan and India in Asia, and the USA and the 28 countries of the EU in the West. Table 5.9 shows figures for total rubber consumption by the three production sectors in 2012 when the total amount of natural rubber and synthetic rubber used as raw material input by the rubber manufacturing industries was 699,070 metric tons. This figure is made up of 441,000 metric tons of natural rubber and 258,000 metric tons of synthetic elastomers. The largest user of solid rubbers was the general products sector that consumed 11.5 per cent, with the tyre sector using 8 per cent. The largest consuming sector by a wide margin was the latex products industry that used 80.5 per cent of raw rubber in the form of latex concentrate as natural and synthetic materials. Table 5.10 provides a breakdown between the consumption of natural and synthetic rubbers.

Table 5.9 Consumption by Production Sector: 2012

Tyres	*General products*	*Latex products*	*Total*
Metric tons	*Metric tons*	*Metric tons*	*Metric tons*
56,514	80,189	562,367	699,070

Note: General products include industrial rubber goods, general rubber goods and footwear.
Source: MRB 2013.

Table 5.10 Consumption of Natural Rubber and Synthetic Rubber: 2012

	Dry rubber	*Latex*	*Total*
	Metric tons	*Metric tons*	*Metric tons*
Natural rubber	75,611	365,505	441,116
Synthetic rubber	61,092	196,862	257,954

Source: MRB 2013.

Synthetic Rubber Production

Before 2000, Malaysia did not manufacture any synthetic rubbers although there were a number of factories that reclaimed and recycled limited amounts of solid synthetics for sale to producers of general rubber goods. The dominant latex glove manufacturing sector only produced goods from *Hevea* latex even though there was a demand from the market for medical examination gloves for products made from liquid nitrile rubber. In 2002, Synthomer, a UK-based multinational producer of polymer chemicals set up production facilities to manufacture nitrile latex (acrylonitrile butatiene latex, XNBR) to supply local glove manufacturers with raw material and SBR latex (carboxylated styrene butadiene latex, XSBR) used in the paper and board coating industries. The factory situated on the site of an existing facility in central Johor, where the company produces modified natural rubber latex concentrate under the Revertex trademark, has a capacity of 120,000 metric tons of nitrile and SBR liquid rubbers. Recently, Synthomer has expanded production on a new site at Pasir Gudang port on the Johor Strait to manufacture 70,000 metric tons of SBR latex and 100,000 metric tons of nitrile latex annually (Synthomer 2013).

The Malaysian Synthetic Rubber company was established in 2013 as a joint venture between Japanese and South Korean chemical manufacturers to produce polybutadiene (BR) rubber at a new factory in Pasir Gudang. Butadiene rubber is a general purpose solid synthetic used in tyres and industrial rubber products. The synthetic production technology is supplied by Ube Industries and Mitsubishi of Japan. The South Korean partner, Lotte Chemical Titan has a refinery in Pasir Gudang where it produces olefins and polyolefins. The company supplies butadiene raw material to the synthetic rubber plant. Production of BR synthetic rubber is currently 50,000 metric tons per year with plans to increase capacity to 75,000 metric tons in the future (Lotte Chemical Titan Holdings 2013; Mitsubishi Corporation 2013).

In late 2013, Petronas, the national oil and gas company signed a joint venture agreement with Versalis, a subsidiary of the Italian petroleum products firm, ENI to manufacture synthetic rubber. The joint venture company will build four synthetic rubber production units at the proposed Petronas petroleum refining and petrochemical complex on the Pengerang peninsula in the south of Johor. The synthetic rubber plants will use chemical technology and technical know-how from Versalis (ENI 2013). Details on the proposed types of synthetic rubber are unknown, although the Third Industrial Master Plan has proposals to establish facilities to manufacture ethylene propylene rubber (EPM) and ethylene propylenediene rubber (EPDM).

By the 2020s, Malaysia will be producing in the region of 75,000 metric tons of solid and 300,000 metric tons of liquid synthetic rubbers annually from existing production facilities. In addition, the Petronas/Versalis joint venture will be manufacturing and exporting synthetic rubbers from its four factories.

Dualistic Structure

In an overview of the rubber manufacturing sector, the UNIDO mid-term review of the first Industrial Master Plan commented that the major feature of the industry was the existence of a large number of small and medium, locally owned firms mainly producing for the domestic market. These companies were generally weak in export trade and were unable to compete, in terms of price, quality and specifications as well as after-sales service, with larger enterprises established wholly or jointly by foreign transnational companies (UNIDO 1991: 75). Taylor and Ward (1994a: 153–5) noted the dualistic structure of the sector, with a small number of large firms controlled by multinational enterprises, and a large number of small and medium-sized industries owned by local companies. The large foreign companies operated factories that were technologically advanced because they had access to the latest manufacturing techniques from abroad. Foreign financial control meant that marketing strategy on export sales and procurement policy on the purchase of manufacturing inputs were decided overseas. In contrast, the smaller Malaysian companies employed less advanced technology with generally poor quality control in manufacturing operations. Tham and Mahani (1999: 65) also stated that the technology of manufacturing is dependent on ownership and size of operations. Wholly foreign-owned firms and joint ventures are larger in size and employ more advanced technology than local companies.

Ten years after the publication of the UNIDO mid-term report, the Director General of the Malaysian Rubber Board and the Deputy Chief Executive Officer of the Malaysian Rubber Export Promotion Council, in their papers on the status of the second Industrial Master Plan (Abdul Hamid Sawal 2001; Ong 2001), noted that there was a relationship with ownership, size of operations and industrial technology employed. Wholly foreign-owned and joint venture firms tended to be larger and use more sophisticated technology, sourced overseas from parent companies or foreign partners. On the other hand, local manufacturers generally operated on a smaller scale and utilized off-the-rack manufacturing processes provided by overseas suppliers of industrial machinery.

Linkage Effects

On the question of linkages with other sectors of the economy, UNIDO commented that businesses producing rubber goods had few backward linkages with manufacturers of factory machinery because the industrial machinery sector was dominated by a large number of small and medium firms producing machinery and equipment for primary commodity producers such as tin mining, palm oil processing and the processing of natural rubber (UNIDO 1991: 103). In 2001 there was local manufacture of glove and other dipped goods machinery, even though tyre fabrication machinery as well as moulding and extrusion equipment were imported from overseas (Abdul Hamid Sawal 2001: 112). The chemicals used in the rubber manufacturing industry were still largely imported. The backward linkages were with producers of a limited number of inputs, namely the domestic manufacture of carbon black filler, kaolin filler, zinc oxide, stearic acid and whiting agents such as titanium dioxide and calcium carbonate. Ong (2001: 119) questioned whether the local production of other rubber chemical agents would be cost-effective in view of the fact that chemically complex accelerators and activators are used in relatively small quantities in manufacturing operations. Rubber product manufacturers had forward linkages with local manufacturing industries, including automobile and motor cycle manufacturers and assemblers, the transport equipment industry, footwear producers, the construction industry, and manufacturers of machinery and equipment (UNIDO 1991: 75).

Conclusion

The manufacture of rubber goods began in the 1920s with the production and export of abrasion-resistant conveyor belts and mining equipment, and rubber-soled canvas shoes and slippers. Just before Independence in 1957, there were approximately 30 manufacturing establishments producing industrial goods, bicycle tyres and tubes, and rubber footwear. The number of companies involved in the production of rubber goods rose from about 70 firms in 1970 to 135 businesses in 1985 at the start of the first Industrial Master Plan. The number of rubber manufacturers had expanded to 357 firms by 2005 when the second plan drew to a close, and it currently stands at 349 companies midway through the third plan.

Rubber products manufacturing has been designated a priority sector in all three industrial master plans. Since 1985 export sales of rubber goods increased from RM215 million to RM14.5 billion in 2012 when exports accounted for 91 per cent of the industry's output. The fact that in 2012 the rubber products industry consumed 699,000 metric tons of natural rubber sourced mainly from Malaysian smallholders suggests that the manufacturing sector is fully integrated with local processors of natural rubber who provide the raw material inputs in a vertically integrated rubber industry as envisaged in UNIDO recommendations in the first Industrial Master Plan.

The literature states that there is a dualistic structure to the industry with a large number of Malaysian-owned firms that are small to medium in size, compared to a small number of large enterprises owned by foreigners or in joint ventures with local partners. The larger companies with foreign investment employ advanced industrial technology sourced from parent companies or joint venture partners overseas. On the other hand, local firms use less advanced industrial procedures, typically provided by the overseas manufacturers of factory machinery. The industry has few backward linkages with Malaysian engineering firms that make industrial machinery although there is local manufacture of machinery for dipped latex goods. There is local supply of a limited number of compounding ingredients including, importantly, carbon black filler. Most rubber chemical agents are, however, imported from overseas. Rubber manufacturers export the greater part of their industrial output. The industry, nevertheless, has strong forward linkages with manufacturers of other products. The sector supplies tyres and automotive components to Malaysian car manufacturers and assemblers of imported cars and motor cycles, and industrial parts and components to local manufacturers of intermediate and consumer goods.

An examination of the literature about the development of the rubber industries sector and the contribution that the sector has made to the process of industrialization from 1970 to the present day indicates that there is a dearth of information compared to the palm oil and electronic sectors. It is suggested that there are two main factors that have influenced economists for their lack of interest in the rubber manufacturing industrial sector as a topic of research. Firstly, the perception of policy-makers and leaders of the plantation industry that natural rubber production, therefore, the manufacture of rubber goods, was a mature industry incapable of further development. An alternative, more pessimistic view widely held during the 70s and 80s was that rubber tree crop agriculture was a sunset industry destined for eventual decline within Malaysia. The lack of confidence regarding *Hevea* rubber production was strengthened by the belief that oil palm was the commodity crop of the future. The 1970s and 1980s were a time of rapid expansion of oil palm culture as the area planted to oil palm in Peninsular Malaysia increased from 260,903 hectares in 1970 to 1.7 million hectares in 1990. Oil palm was planted to the exclusion of other perennial crops on new land by rural development agencies such as the Federal Land Development Authority, while in the private sector the large plantation companies converted vast acreages of old rubber due for replanting to more profitable and less labour-intensive oil palm. The second significant factor was the success of Free Trade Zones in attracting multinational firms to set up operations in Malaysia so that by 1987 exports of manufactured products surpassed the value of the major primary commodities for the first time. However, as a 1989 World Bank report noted, the export manufacturing sector was dominated by transnational and foreign companies located in Free Trade Zones. Thus, commentators on Malaysia's industrialization policy concentrated their investigations almost exclusively on the expansion of export business generated in free trade enclaves, especially the development of the electronics sector, whereas agricultural economists focused their attention on the burgeoning palm oil sector.

This review of the rubber manufacturing industry has revealed that, on the whole, micro-economists have taken only a passing interest in the rubber goods production sector in their studies of the development of industrialization in Malaysia, regarding the sector as being merely one among many in the secondary manufacturing industries. There is thus scant information in the literature on the status of the industry at a micro-economic level and individual firms within it. The release in August 2006 of the *Rubber Industry and Products Directory: 2006–2007* by the Malaysian Rubber Board was the first comprehensive publication on companies manufacturing rubber goods. It also provided a unique record of the rubber manufacturing sector at the end of the second Industrial Master Plan in 2005. At a wider level the directory brought together for the first time a wealth of information on the rubber manufacturing industry that may be tapped as a data base for a micro-economic study on this sector of the larger rubber production industry. Analysis of the data on rubber manufacturing firms in the MRB directory forms the basis for the discussion from Chapter 6 onwards.

CHAPTER 6

Overview of Rubber Manufacturing Industries: 2005

Introduction

The current chapter provides an overview of the rubber products manufacturing industry from an analysis of information published in the trade directory of the Malaysian Rubber Board, *Malaysian Rubber Industry and Products Directory: 2006–2007*, with additional material gathered from the Malaysian Rubber Products Manufacturers' Association's *Industry and Export Directory: 2006–2007*. The two directories provide a detailed record of the rubber manufacturing sector in 2005 at the completion of the Second Industrial Master Plan.

Location of Rubber Industry Factories

The offices and factories of rubber goods manufacturing companies in Peninsular Malaysia are located almost exclusively in the well populated west coast states where there is a developed industrial infrastructure and an extensive road and rail network with easy access to major seaports as shown in the frontispiece map. Table 6.1 shows the location of rubber manufacturing facilities from entries in the MRB directory.

Selangor and the Federal Territory of Kuala Lumpur have the largest concentration with a total of 161 companies compared with other west coast states. Other important centres of production are Perak with 55, Penang with 33 and Johor with 36 rubber manufacturers. The state of Kedah bordering the rubber producing provinces of southern Thailand has 29 rubber manufacturers and is a major producer of rubber gloves and medical products using imports of Thai latex as a factory input.

Table 6.1 Location of Rubber Manufacturing Factories

State	*Number*
Johor	36
Malacca	14
Negeri Sembilan	13
Selangor/Kuala Lumpur	161
Perak	55
Penang	33
Kedah	39
Perlis	2
Kelantan	2
Pahang	1
Total	346

Classification of Rubber Industries

The point has been made in Chapter 2 that rubber manufacturing industries are classified into three major sectors that reflect the manufacturing operations and raw materials employed. These are tyre makers, latex goods manufacturers and producers of general rubber goods. This classification system is regarded as too wide for the purpose of the detailed analysis of the rubber manufacturing industries undertaken in this study. The system used here is to divide the industry into four main product sectors: tyres, latex products, general goods and an additional class for the compounding and retreading materials sector. There are further subdivisions in the tyre, latex and general rubber sectors as shown below:

Tyre industry

- pneumatic tyres
- motorcycle and bicycle tyres, inner tubes and solid tyres

Latex products industry

- latex dipped products
- latex other products

General rubber products industry

- industrial products
- low technology general products
- footwear

Compounding and retreading materials industry

Tyre Industry: *Pneumatic tyre sector*—The sector manufactures automobile tyres for all types of motor vehicles. These include tyres for passenger cars, vans and light trucks, heavy commercial vehicles and agricultural tractors.

Tyre Industry: *Motorcycle/bicycle/solid tyre and inner tube sector*—The products made by this sector include tyres for motorcycles and bicycles, solid rubber tyres for wheelbarrows and warehouse handling equipment, and castors used on domestic furniture. Inner tubes are manufactured for fitting to crossply tyres that dominate the market in less developed parts of the world where road surfaces are typically poor, and overloading of cars, buses and trucks is commonplace.

Latex Products Industry: *Latex dipped products sector*—This sector manufactures dipped goods from latex concentrate to produce general consumer and specialist medical products, such as gloves, condoms, catheters, balloons and swimming caps.

Latex Products Industry: *Latex other products sector*—The items produced by the latex other products sector include extruded latex thread, foam mattress and upholstery products, medical latex sheets, dental and oral dams, latex exercise bands, tourniquet straps and latex toys.

General Rubber Products Industry: *Industrial products sector*—The industrial products sector manufactures a large range of intermediate components for use in the automobile, motorcycle, railway, aircraft and shipping industries. It also makes parts used in the construction and building industry, and in the manufacture of engineering and other industrial products. A defining characteristic of the sector is that it produces intermediate goods used in the manufacture or construction of other manufactured products, infrastructure installations and buildings. The sector does not produce consumer items for sale to the general public.

General Rubber Products Industry: *Low technology sector*—The companies classified as the low technology sector produce a diverse range of consumer and intermediate goods. Items include pencil erasers, bathroom, door and floor mats, elastic bands and industrial straps, sports and play balls, golf balls, tiles and floor covering, adhesives and sealants, rubber sheeting, carpet underlay, swim fins and toys.

General Rubber Products Industry: *Footwear sector*—The firms in the footwear category produce rubber-soled shoes and sandals, sports shoes, industrial safety boots and waterproof boots, as well as sole and heel units used by shoe repairers and manufacturers of boots and shoes.

Compounding and Retreading Materials Industry: The sector makes carbon black masterbatches, specialist masterbatches, and ready-made compounds for sale to small to medium-sized producers of intermediate and consumer goods in other sectors of the rubber manufacturing industry. Products in this sector also include a range of retreading materials for the tyre retreading industry, such as precured tread and sidewall veneers.

Table 6.2 Manufacturers by Production Sector

	Product sector	*No. of companies*
Tyre industry	Pneumatic tyres	3
	Motor cycle/bicycle/solid tyres/inner tubes	13
	Industry total	16
Latex products industry	Latex dipped goods	111
	Latex other products	16
	Industry total	127
General rubber products industry	Industrial products	110
	Low tech general products	47
	Footwear	19
	Industry total	197
Compounding and retreading industry		21
Total		340

Table 6.2 shows the number of firms in each of the eight product sectors. All in all, a total of 340 companies are recorded in the MRB and MRPMA trade directories under the eight categories described above. It should be noted that in the analysis each company has been allocated to a specific sector. However, a fair number of businesses produce more than one type of product so that a calculated judgement had to be made as to which class to place each firm. This has been done in order to avoid double counting in the survey.

Overview of Industry

The data published in the two directories have been collated and analyzed in order to examine the ownership, size of enterprise and export sales in the eight production sectors that make up the rubber manufacturing industry.

Ownership

Table 6.3 gives a breakdown by ownership of the 340 manufacturing companies in the survey. The directory entries do not directly provide this information although the websites of many companies, both those with foreign direct investment and Malaysian-owned, do. However, foreign ownership can be discerned, in many cases, by the name of the business enterprise, for example, Gummi Metall Technik or Nippon Wiper Blade, and by the names of the Chief Executive Officer and other key technical personnel. Typically, companies that are subsidiaries of companies based overseas employ expatriate staff in their Malaysian operations. In the survey of secondary data, it has not proved possible to divide the companies into two separate categories of wholly owned foreign firms and joint venture enterprises.

Table 6.3 Ownership of Companies

Sector	*Ownership of companies*			
	Malaysian		*Foreign/Joint venture*	
	Number	%	*Number*	%
Pneumatic tyres	1	33.3	2	66.6
Motor cycle tyres/inner tubes	12	92	1	8
Dipped goods	88	79	23	21
Latex other products	13	81	3	19
Industrial products	78	71	32	29
Low tech products	38	81	9	19
Footwear	17	90	2	10
Compounds/retread materials	20	95	1	5
Total	267	79	73	21

The figures reveal that the rubber manufacturing industry is mainly Malaysian-owned and operated, accounting for approximately 80 per cent of the industry. However, putting to one side the special case of the pneumatic automobile tyre sector, there are differences among the other seven classes of company. Foreign investment is greatest in the industrial products sector (29 per cent), followed by dipped goods (21 per cent), low technology products (19 per cent) and latex other products (19 per cent) sectors. Local firms account for 90 per cent or more of companies in the footwear, compound/retread materials, and motorcycle tyre/inner tube classes of company.

Among the Malaysian-owned companies, an analysis was carried out to determine the number of privately owned businesses in the hands of different ethnic groups. Out of 267 local companies it was established that 218 or 81 per cent are controlled by private Chinese interests. Only 12 firms are owned by private Malay investors and seven firms by ethnic Indians. The other category of ownership is corporate enterprises, including companies listed on the Malaysian stock exchange, wholly owned subsidiaries of large public listed companies and firms established by quasi-government bodies. Table 6.4 shows the breakdown by ownership of local companies.

Table 6.4 Ownership of Local Companies

Control of capital	*No. of companies*
Private Chinese	218
Private Malay	12
Private Indian	7
Corporate enterprises	30
Total	267

Source of Foreign Direct Investment

Overseas financial investments in Malaysia are typically made by multinational corporations and international manufacturing companies in wholly owned subsidiaries and joint venture enterprises, and this is the case for the rubber manufacturing industry. There is, however, one exception where a Malaysian-controlled producer of latex gloves has direct investment by American and British private equity companies. Japan and South Korea are by far the most important investors, accounting for half of all foreign investment in the rubber manufacturing industry as detailed in Table 6.5.

Investments by Japanese and Korean companies are concentrated in the industrial components sector, reflecting the success of the "Look East" policy adopted when Malaysia was encouraging the establishment of heavy industry in the 1980s. Capital investment from the European Union is in 18 companies across six out of eight sectors, with the exception of the motorcycle tyre/inner tube and compounding/retreading materials industrial sectors. Financial involvement by European and North America manufacturers is found mainly in the latex dipped goods sector, with capital invested in 15 companies. In the footwear sector, investment in a joint venture company by a Norwegian manufacturer of safety boots has been included under the EU heading purely as a matter of convenience. There is also investment in five companies by investors in the Newly Industrialized Economies of Taiwan and Singapore.

Table 6.5 Source of Foreign Investment: Product Sector

Sector	*Source of foreign investment*				
	USA/ Canada	*European Union*	*Japan/ South Korea*	*Australia/ New Zealand*	*Singapore/ Taiwan*
Pneumatic tyres	1	1	0	0	0
Motor cycle tyres/inner tubes	0	0	1	0	0
Dipped goods	7	8	6	2	0
Latex other products	1	2	0	0	0
Industrial products	1	4	22	1	4
Low tech products	0	2	6	0	1
Footwear	1	1	0	0	0
Compounds/retread materials	0	0	1	0	0
Total	11	18	36	3	5

Date of Establishment

As shown in Table 6.6, over 200 firms have been incorporated since 1985 after the formulation by UNIDO of a strategy for the development of

Table 6.6 Date of Establishment: Product Sector

Sector	*Date of establishment*		
	Before 1970	*1970–84*	*1985–2005*
Pneumatic tyres	1	1	1
Motor cycle tyres/inner tubes	2	5	6
Dipped goods	0	14	86
Latex other products	2	4	9
Industrial products	5	30	68
Low tech products	1	12	25
Footwear	2	6	7
Compounds/retread materials	4	3	10
Total	17	75	212

industrialization in the country. The figures demonstrate the success of the policies adopted by Malaysia in attracting direct foreign investment into rubber manufacturing industries as recommended in the first and second industrial master plans. The dramatic expansion of 86 new businesses in the dipped products sector after 1985 was a result of the massive increase in demand for medical examination gloves caused by the spread of AIDS in the 1980s. It is interesting to note that there are still 17 firms in existence that were founded in colonial and immediate post-colonial years that indicates that their management has successfully adapted to changes in the political and business environment over time.

Capitalization

Table 6.7 shows the amount of paid-up capital for the eight classes of company. As discussed in Chapter 3 on manufacturing technology, pneumatic tyre production is undertaken in large, capital-intensive factories where economies of scale operate. The three firms in this sector are, as may be expected, very heavily capitalized in comparison with other sectors. The mean paid-up capital for the pneumatic tyre manufacturing companies is almost RM150 million, whereas among the other classes of manufacturer the mean capitalization is less than RM20 million at the highest. The latex dipped goods and latex other products sectors are the next in importance to the tyre sector, with a mean capitalization of RM17 million each. The remaining manufacturing sectors all have a mean paid-up investment of less than RM10 million.

Table 6.7 Capitalization: Product Sector

Sector	*Capitalization (million RM)*			
	n	*Minimum*	*Maximum*	*Mean*
Pneumatic tyres	3	10.8	231.5	147.5
Motor cycle tyres/inner tubes	12	0.1	30.0	6.3
Dipped products	92	0.2	385.0	17.1
Latex other products	11	0.4	84.0	17.3
Industrial products	81	0.1	51.7	4.1
Low tech products	28	0.1	22.5	2.7
Footwear	6	0.6	8.0	3.6
Compounds/retread materials	18	0.1	13.0	4.2

Workers Employed

The automobile tyre manufacturing sector operates in large-scale production facilities employing a substantial labour force. As shown in Table 6.8, the pneumatic tyre sector is the largest class of employer with a mean employment figure of 1,608 workers per company. The mean employment figure for firms making latex dipped products is 490 workers, and for those making other products from latex is 173 workers per company. Manufacturing concerns making motorcycle tyres and inner tubes, intermediate products for the automotive, engineering and industrial sectors, footwear, and compounding and retreading products employ, on average, more than 100 workers. Only the low technology general goods sector has a mean workforce figure of less than 100 employees.

Table 6.8 Number of Workers: Product Sector

Sector	*Number of workers*			
	n	*Minimum*	*Maximum*	*Mean*
Pneumatic tyres	3	730	3,054	1,608
Motor cycle tyres/inner tubes	13	35	550	133
Dipped goods	102	20	6,500	490
Latex other products	14	16	650	173
Industrial products	90	10	1,160	135
Low tech products	38	10	450	86
Footwear	14	15	510	136
Compounds/retread materials	18	12	400	110

Large and Small/Medium Enterprises

The classification system used to differentiate companies into large and small/medium-scale enterprises (SME) is based on a combination of the amount of capital investment and number of employees. The definition of a large-scale enterprise is one which has over RM2.5 million paid-up capital and over 200 workers. It is also one with either a paid-up capital of over RM2.5 million or having over 200 workers, whichever is the greater. Small and medium-sized firms have less than RM2.5 million paid-up capital and less than 200 employees. One hundred and seventy-two firms are classified as small to medium-scale enterprises, compared to 137 companies in the large-scale class. However, Table 6.9 demonstrates substantial differences among the eight categories of company. The majority of businesses in the motorcycle and inner tube, low technology goods, industrial intermediate products and footwear sectors are classified as small and medium-sized. Only the pneumatic automobile tyre and dipped latex product sectors have a majority of large companies. The three tyre makers are all large-scale companies and 64 firms in the dipped goods sector are classed as large. The figures for the sectors making latex other products and compounds and retreading materials are inconclusive with almost equal numbers in each class.

Table 6.9 Large and Small/Medium-scale Enterprises: Product Sector

Sector	*Large and small/medium-scale enterprises*					
	Large-scale		*Small/medium scale*		*No record*	
	No.	*%*	*No.*	*%*	*No.*	*%*
Pneumatic tyres	3	100	0	–	0	–
Motor cycle tyres/inner tubes	4	31	9	69	0	–
Dipped products	64	58	42	38	5	4
Latex other products	7	44	8	50	1	6
Industrial products	36	33	62	56	12	11
Low tech products	10	21	31	66	6	13
Footwear	4	21	10	53	5	26
Compounds/retread materials	9	43	10	48	2	9
Total	137	40	172	50	31	10

Export Sales

Examination of the data on export sales reveals that 77 firms in the latex dipped products sector export more than half of factory output, indicating that this class of manufacturer produces largely for the export market. A majority of business concerns in the industrial components sector, low technology general goods and other latex products classes also sell more than 50 per cent of production overseas. Those sectors that produce mainly for the domestic market are pneumatic tyres, motorcycle and inner tubes, and footwear, while the data are inconclusive for manufacturers of compounding and retreading materials. When taken as a whole, the rubber manufacturing industry is strongly export-oriented with 134 companies (73 per cent) out of the 184 that provided information selling more than half of their total production into markets in foreign countries.

There are differences in export sales strategy among individual sectors of the rubber products industry. Although the pneumatic tyre sector exports less than 50 per cent of production, the industry is strongly export-oriented with sales across the globe. The most important market for the dipped goods class of producer is the European Union, with 99 out of 105 exporters selling into this market. The Association of Southeast Asian Nations (ASEAN) is the most significant overseas market for the industrial products sector that makes automotive and engineering components. Out of 86 firms that reported export sales, 73 businesses supply nearby markets in the ASEAN regional grouping. Markets in the less developed regions of the world are of greater importance to the other five production sectors compared to markets in industrialized economies. All 15 exporters of compounding and retreading products sell into developing countries as do the majority of exporting firms in other sectors.

Data published in the trade directories demonstrate that Malaysian-based manufacturers sell into markets across the globe and that the rubber manufacturing industry, when taken as a whole, is not reliant on any one export market. More Malaysian-based exporters, 212 in total, supply manufactured goods to other newly industrializing and developing countries in the "rest of the world" than to markets in the industrialized Far East and the West. The next two important markets, in terms of the number of companies that undertake export activities are the EU and ASEAN. A marginally smaller number of exporters sell into the North American, Japanese/South Korean and Australian/New Zealand markets. The overall conclusion is that export sales are widely based and that South-South trade is important to Malaysia.

Conclusion

Analysis of data recorded by 340 manufacturing firms in the eight production sectors in the survey shows that the rubber products manufacturing industry is highly differentiated in terms of ownership, size of enterprise and export sales. However, when the industry is taken in the round some common trends may be discerned. Some 80 per cent of firms are Malaysian-owned and operated. The majority of local companies (approximately 80 per cent) are owned by ethnic Chinese private investors. Only a small number of enterprises are owned by private Malay and Indian interests, or are controlled by public listed companies. Companies with overseas capital involvement, either as wholly owned subsidiaries of foreign companies or in joint venture with Malaysian interests, account for 20 per cent of the industry. The Far Eastern industrialized economies of Japan and South Korea are jointly the most important source of foreign investment, followed by countries of the European Union. Direct foreign investment is greatest in the automotive, engineering and industrial sector with almost 30 per cent of companies having overseas involvement.

Most companies were established after 1985 when the First Industrial Master Plan became operational, although a small number of manufacturing concerns can trace their roots to colonial times and the early years of Independence.

The majority of businesses are classified as small to medium-sized enterprises although there are differences among the production sectors. The three automobile tyre makers are substantial businesses by any standards and the majority of the large companies are in the latex dipped products sector.

The industry is strongly export-oriented. The majority of firms, 134 in total, export more than 50 per cent of factory production, compared to 43 firms that sell less than half of their output overseas. Only seven companies reported that they produce solely for the domestic market. Information on export markets was provided by 273 enterprises, or 80 per cent of companies in the analysis. Export sales are widely based and the industry is not overly dependent on any single market. The most significant export market is not in the industrialized countries but in the developing economies of the "rest of the world" and neighbouring countries in ASEAN. The evidence is that South-South trade is important to Malaysian exporters. A greater number of firms export to the EU than to other developed economies in USA/Canada, Japan/South Korea and Australia/New Zealand. However, these figures do not reflect the value of exports that are sold into each market although the Malaysian Rubber Export Promotion Council indicates that the two largest markets in terms of export revenues are the USA and the European Union.

The following two chapters extend the micro-economic analysis in depth by examining each sector in more detail and making a comparison between wholly Malaysian-owned companies on the one hand, and foreign and joint venture business enterprises on the other. Chapter 7 considers the case of the automotive, engineering and industrial component industry while Chapter 8 has a more cursory analysis of the other production sectors.

CHAPTER 7

Automotive, Engineering and Industrial Components Industry

Introduction

Chapter 7 undertakes an in-depth examination of manufacturing companies producing intermediate components for the automotive, engineering and other manufacturing industries. It analyses secondary data from directories published by the Malaysian Rubber Board and Malaysian Rubber Products Manufacturers' Association for 2006, and primary data collected by means of a questionnaire survey. The focus of the analysis is on a comparison between manufacturers with direct foreign investment and those wholly owned by Malaysian business interests.

The sector is large, with 110 firms making industrial components and intermediate products for supply to manufacturing companies in the automotive, mining, civil construction, engineering and other production industries. Tables 7.1 and 7.2 show the range of products manufactured by companies in the sector. It should be noted that while most firms manufacture a range of products, a certain number specializes in the production of a particular product.

Analysis of Trade Directories

Ownership

There is a strong overseas presence, with 32 business enterprises (29 per cent) being either fully foreign-owned or in joint venture with Malaysian interests out of the total of 110 manufacturing concerns. The industrial

Table 7.1 Automotive Components

Products	*No. of companies*
Automobile bushings	28
Automobile seals	24
Bumpers	5
Car mats	17
Engine mountings	13
Radiator hoses	5
Steering driveshaft boots	7
Weatherstrip	7
Wiper blades	3
Motor cycle parts	8
Railway components	8

Table 7.2 Engineering and Industrial Components

Products	*No. of companies*
Rubber to metal bonded parts	29
High precision products	5
Construction/civil engineering products	19
Electrical/electronic products	55
Mining industry parts	13
Anti-vibration mounts	10
Belting	4
Bridge bearings	9
Dock fenders	8
Hoses	19
Insulated cables	2
Seals and gaskets	36
Rollers, industrial	12
Rollers, printing	11
Tank lining	10
Water stops	5

components sector has a greater degree of foreign participation compared to the other production sectors.

Source of Foreign Investment

The most significant source of overseas capital is Japan by far. It has investments in 22 businesses, with other countries being of lesser importance, as shown in Table 7.3.

Table 7.3 Source of Foreign Investment: Industrial Sector

Country	*No. of companies*
Japan	22
European Union	4
USA	1
Australia	1
Singapore	2
Taiwan	2

It is significant that Malaysia has attracted inward investment from two newly industrialized economies: Singapore and Taiwan. Four EU countries have investments in the industrial components sector, namely Denmark, Germany, Italy and the United Kingdom. Details of these companies are provided in Table 7.4 as an example of the types of specialist products manufactured by the foreign and joint venture sub-sector. Alfagomma-Mardec and Pong Codan are joint venture enterprises between Malaysian interests and the Italian transnational Alfagomma and the Danish-owned Codan Gummi company respectively. Gummi Metall Technik and Linatex are wholly owned subsidiaries of German and British international manufacturers.

Table 7.4 European Union Investments

Company	*Source of foreign investment*	*Main products*
Alfagomma-Mardec	Italy	Hydraulic hoses
Gummi Metall Technik	Germany	Anti-vibration rubber to metal bonded products
Linatex Rubber Products	UK	Abrasion and corrosion resistant lining and sheeting, mining pumps
Pong Codan	Denmark	Moulded automotive components, extruded weatherstrip, hoses

Date of Establishment

Two firms, the Malaysian-owned Shum Yip Leong Rubber Works and the British-controlled Linatex Rubber Products, were established in the 1920s. Another three local firms, Kinta Rubber Works, Swan Rubber Products and Universal Cable were founded before 1970. Between 1970 and 1984, under the New Economic Policy, 28 Malaysian businesses began operations, compared to only two firms with overseas financial involvement, the German-owned, Gummi Metall Technik and Kee Fatt Industries, a Malaysian-Japanese joint venture with majority investment by the local firm. As indicated in Table 7.5, the majority of the companies were incorporated after the implementation of the Industrial Master Plan in 1986. These are made up of 39 (50 per cent) of the 78 Malaysian firms and 29 (90 per cent) of the 32 businesses with overseas interests. The latter includes 21 firms with investment from Japan.

Table 7.5 Year of Incorporation

Ownership	*before 1970*	*1970–84*	*1985–2005*
Malaysian	4	28	39
Foreign/joint venture	1	2	29

Capitalization

A single company, Kossan Rubber Industries is quoted on the main board of the Malaysian Stock Exchange. This firm, incorporated in 1979, has RM51.7 million paid-up shares and employs 390 workers. It produces moulded and extruded intermediate goods for automotive, engineering, industrial, construction, mining and marine applications. The company has two subsidiary associate companies. Kossan Japan Rollers is a joint venture with a Japanese partner that manufactures rollers for office equipment. The wholly Malaysian-owned Kossan Latex Industries makes examination gloves for the medical and healthcare professions.

The publicly listed Kossan Rubber Industries with over RM50 million paid-up share capital is, as may be expected, much more heavily capitalized than companies that do not raise investment finance on the stock exchange. The next largest Malaysian enterprise in term of capitalization is the Jebco Group with an investment of RM20 million, less than half the amount of paid-up capital compared to Kossan. Kossan Rubber Industries as a "statistical outlier" company, therefore, has been removed from the analysis of a comparison of capital investment in Malaysian-owned firms and those companies with foreign capital involvement. Table 7.6 shows the figures for paid-up capital in each class of company. Statistical calculations indicate that local businesses with a mean figure of RM2.2 million have significantly less investment than foreign and joint venture enterprises with a mean of RM6.1 million paid-up share capital.

Table 7.6 Capitalization (RM million): Industrial Sector

Ownership	*n*	*Minimum*	*Maximum*	*Mean*
Malaysian	53	0.1	20	2.2
Foreign/joint venture	27	0.3	26	6.1

Note: Kossan Rubber Industries removed.

When an examination of the ten most heavily capitalized firms (defined as those enterprises with a capitalization of RM10 million or more) was made, the results show that four local businesses fall into this classification compared to six companies with investment from overseas. Of the four local companies Seginiaga Rubber Industries, a manufacturer of weatherstrip and window seals, and the Jebco Group, a producer of moulded components and bonded rubber to metal parts, are private companies. The other local companies raise capital on the stock exchange, directly in the case of Kossan Rubber Industries, and indirectly in the case of Mardec Polymers. The latter company was established in 2001 as a downstream manufacturing subsidiary of the publicly listed Mardec to make consumer and industrial products. The principal business of Mardec, a former government agency, is to process smallholder rubber in a chain of factories located in the rubber growing areas of the country.

Mardec is also the local partner with an Italian manufacturer in a joint venture enterprise, Alfagomma-Mardec that produces hydraulic hoses. Capital from Japan is invested in two joint venture companies, Marutech Elastomer and Technomeiji, with a Malaysian manufacturing group to supply automotive components to local car manufacturing industries. Kozato Kizai and Nichias FGS are wholly owned subsidiaries of Japan-based companies. Kozato Kisai makes components for the electrical/electronic sector, and Nichias produces gaskets for the petroleum extractive and automotive industries. Linatex, as has been noted previously, is a British rubber manufacturer that specializes in the production of abrasion and corrosion-resistant products. Table 7.7 shows the details.

Workers Employed

A statistical analysis of the size of the workforce employed in local firms and in foreign and joint venture companies reveals a picture similar to the degree of capitalization in the two sectors. Malaysian companies are smaller in size as compared to companies with foreign involvement. Companies with overseas capital employ a significantly greater number

Table 7.7 Heavily Capitalized Firms

	Company	*Paid-up capital*	*Source of investment*
Malaysian Firms	Seginiaga Rubber Industries	12.5	Private
	Mardec Polymers	13.5	Stock market via Mardec
	Jebco Group	20.0	Private
	Kossan Rubber Industries	51.7	Stock market
Foreign/ Joint Venture Firms	Alfagomma-Mardec	10.0	Italy/Malaysia
	Marutech Elastomer	10.0	Japan/Malaysia
	Kozato Kizai	11.2	Japan
	Technomeiji	12.5	Japan/Malaysia
	Linatex Rubber Products	15.9	UK
	Nichias FGS	26.0	Japan

Note: RM 10 million and more paid-up capital.

of workers than local enterprises. The mean figure for Malaysian-owned businesses is 88 employees, compared to 234 workers in joint venture and foreign firms. Table 7.8 has the figures.

Table 7.8 Number of Workers: Industrial Sector

Ownership	*n*	*Minimum*	*Maximum*	*Mean*
Malaysian	61	10	520	88
Foreign/joint venture	29	30	1160	234

Table 7.9 Large Employers

Company	*Workforce*	*Products*
Universal Cable	520	Insulated cables
Inoac FKR	650	Office equipment rollers
Nichias FGS	720	Industrial gaskets
Nippon Wiper Blade	800	Automobile wiper blades
Yamauchi Malaysia	1160	Electrical/electronic parts

Note: Over 500 workers.

As shown in Table 7.9, a total of five companies employ a labour force of 500 or more workers: one Malaysian business and four firms with Japanese shareholders. The insulated cable and wire manufacturer, Universal Cable, formerly quoted on the Kuala Lumpur Stock Exchange but now delisted, has a workforce of 520. Inoac FKR is a joint venture between the Inoac Corporation of Japan and motorcycle tyre maker Fung Keong Rubber Manufactory. Nippon Wiper Blade is an offshore production facility of the eponymous company in Japan. The Malaysian subsidiary of Yamauchi Corporation makes components for the electrical and electronics manufacturing sector and is the largest employer of labour. However, the largest company in terms of the amount of capital and size of the labour force is the Japanese-controlled Nichias FGS, with RM26 million paid-up capital and 720 workers.

Large and Small/Medium Enterprises

A comparison of the number of large and small/medium-scale enterprises in each sector indicates that companies that employ capital from overseas are significantly larger in size (over RM2.5 million and/or over 200 workers) than business enterprises wholly owned by Malaysian investors. Table 7.10 shows 75 per cent of foreign and joint venture companies fall under the large-scale category, compared to a mere 15 per cent of locally owned firms, whereas 70 per cent of local businesses are classified as small and medium-sized enterprises.

Table 7.10 Large and Small/Medium-scale Enterprises: Industrial Sector

Ownership	*Large-scale*		*SME*		*No record*	
	No.	*%*	*No.*	*%*	*No.*	*%*
Malaysian	12	15.4	55	70.5	11	13.1
Foreign/joint venture	24	75.0	7	21.9	1	3.1

Volume of Export Sales

The directories provide data on the volume of production exported to overseas markets for 34 local businesses and 21 firms employing foreign investment capital. In percentage terms these figures translate as 44 per cent of Malaysian companies and 65 per cent of joint venture and foreign-owned enterprises which give details on export sales as a percentage of factory output. Examination of the figures reveals that foreign and joint venture companies export a larger volume of production than local businesses. Thus, of those companies that undertake overseas sales, 90 per cent of firms with foreign share capital export between 50 and 100 per cent of their factory production, whereas the figure for Malaysian firms is 38 per cent of the total. Three local businesses reported that they have no export sales, and 21 firms export less than 50 per cent of total output. The conclusion is that, on the whole, locally owned firms concentrate on the supply of industrial components to manufacturing industries based in Malaysia whereas firms with foreign capital are typically more export-oriented.

Export Markets

Information on sales into individual export markets was provided by 58 local manufacturers (74 per cent) and 28 companies (87 per cent) employing overseas capital. No figures were given by 20 local and 4 foreign and joint venture enterprises. As has been noted above, three companies owned by Malaysian interests sell exclusively into the domestic market. The most important market for the supply of made-in-Malaysia industrial parts and components is the ASEAN market. Over 80 per cent of the companies, both local and foreign, export to the region. Second in terms of sales is "the rest of the world" with two-thirds of all manufacturing companies selling into newly industrializing and developing countries. The predominance of Japanese direct foreign investment in the sector is reflected in the figure for sales into the Japanese and South Korean market, with 60 per cent of foreign and joint venture companies supplying these two countries. Malaysian exporters are in a stronger position than companies with foreign finance for sales of products into the industrial markets of the European Union, Australia and New Zealand and North America. Table 7.11 provides details of export markets supplied by the sector.

Table 7.11 Export Markets: Industrial Sector

Ownership	*Markets*					
	USA/ Canada	*European Union*	*Japan/ South Korea*	*Australia/ New Zealand*	*ASEAN*	*Rest of the world*
Malaysian	28	32	24	30	50	39
Foreign/joint venture	10	9	17	11	23	18

Analysis of Primary Data

A postal survey was undertaken by means of a questionnaire. Questions included topics such as ownership, source of investment funds, size of enterprise and export sales. There were also questions on linkages with suppliers of factory inputs as well as the source of production technology and technical advice. The purpose of the last set of enquiries was to determine the impact of industrial technology developed by the MRB compared to technology transferred from overseas by parent companies or joint venture partners. Out of a total of 82 questionnaires, 25 were returned and the results of the survey are presented in this section of the chapter.

Ownership

A total of 17 firms (68 per cent) are wholly owned by Malaysian interests, and 8 firms are either joint ventures or wholly foreign-owned. There are two joint venture companies with a majority Malaysian and minority foreign financial stake; and three joint ventures with majority foreign and minority Malaysian ownership. Three foreign-owned businesses are wholly held subsidiaries of overseas manufacturers.

Source of Foreign Investment

The companies with overseas involvement appear to be representative of the sector as a whole. Answers to the questionnaire show that out of the eight businesses employing foreign capital, two have investments from Japan, whereas capital from the USA, EU, Australia, Taiwan and Singapore is invested in one company each. The source of foreign investment and the type of company in the joint venture and foreign-owned companies is shown in Table 7.12.

Table 7.12 Source of Foreign Investment and Type of Enterprise

Type of company	*Number of companies*	*Source of foreign investment*
Joint venture: majority Malaysian-owned	1	Japan
	1	USA
Joint venture: majority foreign-owned	1	European Union
	1	Singapore
	1	Taiwan
Foreign-owned	2	Japan
	1	Australia

Large and Small/Medium Enterprises

Table 7.13 demonstrates that 62.5 per cent of joint venture and foreign-owned enterprises are in the large-scale industry category, compared to

35 per cent of Malaysian companies. The situation is reversed when observing small and medium-scale manufacturing firms: 65 per cent are Malaysian-owned and 38 per cent have foreign involvement. The result supports the conclusion of the analysis of the secondary data that companies with direct foreign investment, either as wholly owned or joint venture business enterprises are likely to be larger than companies employing only Malaysian capital.

Table 7.13 Large and Small/Medium-scale Enterprises: Postal Survey

Ownership	*Size*	*Frequency*	*Per cent*
Malaysian	Large-scale	6	35.3
	SME	11	64.7
Foreign/joint venture	Large-scale	5	62.5
	SME	3	37.5

Volume of Export Sales

Examination of the questionnaires reveals that three companies are in Free Trade Zones. Free Trade Zones are tax-free enclaves that provide incentives for foreign-owned, export-oriented companies to locate their production facilities. One company is a capital-intensive, majority foreign/minority Malaysian joint venture with investment by a major, European-based, multinational manufacturer. The other two companies are wholly owned subsidiaries of Japanese manufacturers with production facilities in Selangor. Looking at the volume of exports from the three companies located in Free Trade Zones, two manufacturers, namely the joint venture company and one of the Japanese firms, export up to 100 per cent of production, whereas the other Japanese company exports between 50 and 74 per cent of its output. Of the joint venture and foreign-owned manufacturers, 75 per cent sell over half of their production in overseas markets, in contrast to 35 per cent of Malaysian industrial concerns. On the other hand, almost 60 per cent of Malaysian companies export less than 50 per cent of production and a single company owned by Malaysians sells only into the home market.

These figures clearly demonstrate that, of the companies in the survey, there is a relationship between ownership of manufacturing companies and the amount of production that is exported. Businesses that are foreign-owned or have part of the capital owned by foreign investors export a greater volume of their production compared to their Malaysian counterparts. Once again, the results of the postal survey support the conclusion reached in the examination of the secondary data.

Export Markets

The figures in Table 7.14 throw up some interesting comparisons between the two categories of manufacturers of intermediate industrial goods. Firstly, despite the small size of the sample, both Malaysian producers, and foreign and joint venture industries sell into all six markets. The member countries of the ASEAN community are the most important export market for both categories of producers, as is the case for the automotive, engineering and industrial sector taken as a whole. On the other hand, Malaysian companies sell proportionately more into the industrialized countries of North America and Europe, and Australian/New Zealand markets compared to companies with foreign investment. The proportion of exports sold into the Japanese/South Korean market, and the rest of the world are the same for both classes of manufacturer. The smallest export market for manufacturers of industrial rubber goods based in Malaysia is the rest of the world, which is not the case for the entire rubber manufacturing industry.

Table 7.14 Export Markets: Postal Survey

Ownership	*Markets*					
	USA/ Canada	*European Union*	*Japan/ South Korea*	*Australia/ New Zealand*	*ASEAN*	*Rest of the world*
Malaysian	8	11	8	11	12	4
Foreign/joint venture	3	3	4	2	7	2

Source of Natural Rubber

Table 7.15 indicates that most Malaysian as well as foreign and joint venture producers of industrial goods use only natural rubber produced in Malaysia as raw material. There appears to be very little difference between the two categories, with 5 or 62.5 per cent of companies with overseas investment, and 9 or 53 per cent of Malaysian companies purchasing 100 per cent of Malaysian-produced natural rubber as a manufacturing input. Non-Malaysian sources of natural rubber are: Thailand that supplies raw material to eight companies; Indonesian rubber is an input for two firms; other ASEAN countries, such as Vietnam or Cambodia, supply five enterprises; and two manufacturers use natural rubber from non-ASEAN countries. One joint venture company that sources less than 25 per cent of its raw material from Malaysia purchases its natural rubber from Thailand and other ASEAN countries. The Malaysian firm that buys most of its rubber from overseas indicated that it purchases its raw material from both ASEAN and non-ASEAN sources.

Table 7.15 Procurement of Natural Rubber from Malaysia

Ownership	*Rubber purchases*	*Frequency*
Malaysian	100%	9
	75–99%	5
	50–74%	2
	25–49%	1
	less than 25%	0
Foreign/joint venture	100%	5
	75–99%	2
	50–74%	0
	25–49%	0
	less than 25%	1

Source of Synthetic Rubber

Sixteen Malaysian producers use synthetic rubbers in the manufacturing process, and one small manufacturer uses only natural rubber. On the other hand, all the foreign and joint venture companies include synthetic rubber inputs in their manufacturing operations.

Among the Malaysian companies, 13 of them, presenting slightly over 80 per cent of users of synthetic materials, purchase 50 per cent or more of their synthetic rubbers and other elastomers from suppliers based in Malaysia, compared to three firms that buy most of these factory materials from overseas suppliers. In contrast, three-quarters of manufacturers with foreign involvement purchase 50 per cent or more of their synthetic materials from suppliers based in overseas countries, with three companies out of eight buying all of their synthetic supplies outside Malaysia. The figures demonstrate that there is a difference in purchasing behaviour between Malaysian firms on the one hand, and foreign and joint venture firms on the other. Joint venture and foreign-owned manufacturers buy most of their synthetic inputs from suppliers in overseas countries, compared to their Malaysian counterparts who purchase the majority of their inputs from companies based in Malaysia.

Source of Compounding Ingredients

There is generally a difference in the use and procurement of compounds between larger manufacturing concerns and smaller business enterprises, as described in Chapter 2 on manufacturing technology. Firms that operate on a large scale normally manufacture their own masterbatch and other compounds on site in their own factory facilities. These large manufacturers typically purchase compounding ingredients in order to make custom-made compounds tailored to their own specifications. Smaller companies, on the other hand, purchase masterbatch and other compounds from specialist manufacturers of these products and typically would include other ingredients such as fillers and accelerators in the manufacturing process.

However, bearing this caveat in mind, there is a clear difference in procurement policy between production companies employing overseas capital and those wholly owned by Malaysian capital. Among Malaysian firms, 65 per cent source over half of the compounding materials from producers of compounds and compounding ingredients based in the country. Five firms source their compounding ingredients only from manufacturers with factories located in Malaysia. On the other hand, not a single company with foreign investment uses compounding ingredients made only by Malaysian factories. In contrast to the Malaysian-owned companies, seven out of the eight manufacturers owned by foreign and joint venture interests use compounding ingredients made mainly in overseas countries as production inputs. One firm uses compounding materials purchased exclusively from overseas sources in its factory operations.

When an examination is undertaken to compare the procurement policies between large companies owned by Malaysians and those with overseas involvement, the figures suggest that there is a difference in purchasing behaviour between the two classes of company. Three locally owned companies but only one firm with foreign participation source the greater part of their requirements for compounding ingredients from manufacturers based in Malaysia. By way of contrast, the remaining three Malaysian firms and four out of five joint venture and foreign companies use more compounding materials produced by manufacturers in countries overseas. The results indicate that the difference observed in the purchase of compounding materials is dependent on the ownership of assets rather than the size of the company.

Suppliers of Compounding Ingredients

Malaysian-owned factories purchase the greater proportion of their supplies of masterbatch and manufacturing chemicals from firms based in Malaysia, in contrast to foreign and joint venture manufacturers who source these materials mainly from suppliers outside Malaysia. Among the Malaysian companies, 14 out of 17 buy between 75 to 100 per cent of their compounding inputs from suppliers in the country, whereas 6 out of 8 joint venture and foreign manufacturers buy 50 per cent or more of their compounding supplies from overseas.

Source of Machinery and Equipment

There is no difference between the two categories of company in their purchases of manufacturing machinery. The figures reflect the fact that Malaysia lacks a sophisticated machine tool and engineering manufacturing industry that is able to supply inputs of capital machinery to other manufacturing sectors. Almost all of the manufacturing equipment used by respondents in the survey is imported from overseas manufacturers of industrial machinery, with only a small percentage being made in Malaysia.

Malaysian Rubber Board Technology

The survey clearly demonstrates that Malaysian-owned companies rely on the Malaysian Rubber Board as their source of manufacturing technology, compared to firms that have foreign investment. A total of 14 out of 17 Malaysian-owned industrial concerns indicate that they use manufacturing technology developed by the Malaysian Rubber Board. A joint venture firm with a majority control from Singapore uses MRB technology in contrast to the other companies with foreign participation.

Other Sources of Technology

The importance of the MRB as a source of manufacturing technology for Malaysian manufacturers has been highlighted in the previous paragraph. This class of company also has a heavy dependency on other sources of technology. Firstly, on manufacturing methods developed in-house on the factory floor; secondly, technology that is common knowledge throughout the industry; and thirdly, in laboratories attached to the production facilities. Technology transfer agreements with overseas manufacturers are important to only three Malaysian companies. As might be expected, the majority of joint venture and foreign-owned firms source their manufacturing technology from parent companies based overseas, although for three manufacturers, in-house laboratories are important in developing production technology.

Malaysian Rubber Board Technical Assistance

The response of manufacturers of industrial rubber products in each category of ownership to the question on the source of technical advice mirrors that on production technology. A total of 13 Malaysian companies use the services of the MRB for technical assistance in manufacturing operations and upgrading manufacturing technology. The same joint venture that uses MRB technology contacts the Board for advice on problem-solving in factory process procedures.

Other Sources of Technical Assistance

Both classes of manufacturer use company-owned laboratories as a source of technical advice for problems that arise in the manufacturing process. In-house laboratory facilities are regarded as important to 70 per cent of Malaysian producers, compared to 50 per cent of foreign and joint venture companies. The internet is used for problem-solving in factory operations by both categories of producer. Six companies with foreign investment use advisory services provided by their overseas investors or through a technology licence. In contrast, only three Malaysian companies use a technology agreement licence with an overseas manufacturer for technical assistance. As the answer to the previous research question has shown, Malaysian manufacturers depend heavily on the technical advisory services of the MRB for help in manufacturing operations.

Conclusion

The automotive, engineering and industrial production sector is large, consisting of 110 firms, and has a greater degree of direct foreign investment compared to other sectors of the rubber products manufacturing industry. Companies owned by Malaysian interests comprise 78 business enterprises and overseas capital is invested in 32 manufacturing concerns.

Japan is the most important source of foreign investment. The first firms were established in the 1920s. The majority of companies have been incorporated in the recent 20-year period, 1985–2005.

The analysis of primary and secondary data clearly demonstrates differences in business behaviour between Malaysian-owned companies on the one hand, and joint ventures and foreign companies on the other. There are, however, similarities in the two classes of company in three business areas: purchase of natural rubber inputs; the source of factory machinery; and export sales. Natural rubber produced in Malaysia is purchased by the majority of companies whereas manufacturing equipment and industrial machinery are sourced from overseas manufacturers. The most significant export market for industrial rubber products in terms of the number of exporting firms that supply each market is the ASEAN region.

The dissimilarities between manufacturing enterprises controlled solely by Malaysian capital and industrial concerns with investment by stakeholders in foreign countries are summarized in the concluding statements:

Capitalization: Foreign and joint venture enterprises are significantly more heavily capitalized than local firms, except for the case of the publicly listed Kossan Rubber Industries.

Employment: Companies with overseas investment employ a significantly larger workforce than Malaysian manufacturers.

Size of company: Joint ventures and foreign-owned companies are significantly larger in size, measured by the amount of paid-up capital and number of workers, compared to Malaysian producers.

Volume of exports: Firms employing foreign capital export a greater proportion of production than Malaysian companies.

Synthetic rubber suppliers: The majority of manufacturers with foreign investment buy over half of their inputs of synthetic rubbers from overseas suppliers, whereas the majority of Malaysian producers purchase almost all their synthetics from suppliers based in the country.

Manufacturers of compounding ingredients: Producers of rubber goods owned by foreign and joint venture companies source most of their compounding ingredients from overseas manufacturers. Most Malaysian firms purchase a greater proportion of compounding materials from manufacturers with factories located in Malaysia.

Suppliers of compounding ingredients: The majority of Malaysian rubber product manufacturers obtain a greater proportion of their compounding materials from suppliers based in the country. Joint venture and foreign-owned manufacturers source most of their compounding inputs from overseas suppliers.

Malaysian Rubber Board technology: 80 per cent of Malaysian manufacturers use production technology developed by the Malaysian Rubber Board, compared to a single joint venture enterprise.

Other sources of manufacturing technology: Foreign and joint venture manufacturers typically source manufacturing technology from their overseas partners or parent companies.

Malaysian Rubber Board technical assistance: Malaysian-owned manufacturers of industrial rubber goods rely heavily on the advisory services of the Malaysian Rubber Board in solving problems that arise during manufacturing operations and for advice on upgrading production technology.

Other sources of technical assistance: Most manufacturers with foreign participation obtain technical assistance from their overseas parent company or joint venture partner. Technical advice provided by company-owned laboratories situated on site is important to 70 per cent of Malaysian and 50 per cent of joint venture and foreign-owned companies.

The conclusion to be drawn from this chapter is that the industrial goods sector exhibits a clear dual structure with foreign and joint venture companies being much larger in size than their Malaysian counterparts. Malaysian companies, on the other hand, have closer links with the domestic economy for both the procurement of factory inputs and with the Malaysian Rubber Board as a source of manufacturing technology.

CHAPTER 8

Tyre, Latex, General Products and Compounding Industries

Introduction

The analysis briefly considers such questions as ownership, capitalization, size of workforce and exports for the following production sectors:

- pneumatic tyres;
- motorcycle and bicycle tyres, inner tubes and solid tyres;
- latex dipped products;
- latex other products;
- low technology general products;
- footwear; and
- compounds and retreading materials

Pneumatic Tyre Sector

The companies that manufacture pneumatic tyres are Continental Sime Tyres, a joint venture between the German tyre maker, Continental AG and the Malaysian conglomerate Sime Darby; Goodyear Malaysia, a subsidiary of the American tyre multinational, Goodyear Tire and Rubber Company; and Silverstone, a wholly owned subsidiary company of the public listed Lion Group of Malaysia. Continental Sime was incorporated in 2003 when Sime Darby, that had a controlling interest in Dunlop Malaysia, entered into partnership with the German multinational company. The Dunlop tyre factory began operations in 1961 in newly independent Malaya while Goodyear Malaysia was set up under the NEP in

1972. The local manufacturer, Silverstone was established in 1988 after the First Industrial Master Plan had recommended the expansion of the tyre making sector in the country.

All three manufacturers produce radial and crossply tyres for passenger and commercial vehicles, and tyres for farm tractors that are sold in Malaysia and overseas markets. In addition, Continental Sime makes massive earthmover tyres for heavy equipment used in the construction and mineral excavation industries, whereas Silverstone concentrates on the production of tyres for four-wheel-drive vehicles, and has facilities to produce high performance motor racing and rally tyres.

Table 8.1 Pneumatic Tyre Companies

Company	*Ownership*	*Year established*	*Paid-up capital (RM million)*	*Number of employees*
Continental/Sime	Germany/Malaysia	1961	231.5	3054
Goodyear	USA	1972	10.5	730
Silverstone	Malaysia	1988	200.3	1040

Tyre manufacture is a capital-intensive enterprise undertaken in large automated factories employing hundreds of workers, as illustrated in Table 8.1. Both Continental Sime and Silverstone are highly capitalized with paid-up investments of over RM200 million (US$50 million) each. Goodyear Malaysia has RM10.5 million (less than US$3 million) invested as paid-up capital, which perhaps is a reflection of the desire to keep executive and financial control firmly in the hands of the US parent company. The largest company in terms of investment and employment is the German/Malaysian joint venture that operates on two production sites, compared to the other two firms that have a single factory each. The technology of tyre fabrication is supplied by the German and American parent companies in the case of the two companies with overseas equity, whereas it is reported that Silverstone has a technology agreement with the British specialist tyre manufacturer, Avon.

Besides supplying original equipment tyres to local automobile manufacturers and assemblers, and selling replacement tyres in the domestic market, the Malaysian tyre sector is a major exporter, with sales to countries across the globe, from Australia and islands in the Pacific to Zambia and Zimbabwe in Africa. Two companies, Silverstone and Continental Sime indicate that they export 40 per cent of production; however, no details of volume of exports are provided by Goodyear Malaysia. The entry for Goodyear in the MRB directory shows that the company exports to 20 countries in 5 continents, whereas Continental Sime and Silverstone state that their export market is worldwide without providing further details. Fuller information on export markets was provided in the Federation of Malaysian Manufacturers (FMM) trade directory for 2003. It shows that Goodyear is not as strong in export markets as Silverstone and Continental Sime. This situation is hardly surprising given that the US-controlled Goodyear is part of a multinational company with subsidiaries and production facilities in the majority of markets for tyres. Within ASEAN, for example, Goodyear has long established subsidiary companies in Indonesia, the Philippines and Thailand, so Goodyear Malaysia is constrained in promoting export sales to neighbouring countries. Continental Sime (as the former Dunlop Malaysia) and Silverstone, on the other hand, have been free to pursue an expansionary export strategy and have built a strong international market share, particularly in Europe, the Middle East and sub-Saharan Africa. The performance of Silverstone in building up a worldwide market for a new brand of tyre only 20 years since the company began operations is impressive.

Motorcycle/Bicycle/Solid Tyre and Inner Tube Sector

Out of the 13 firms in this sector, 12 are wholly Malaysian-owned and one manufacturer of castors, NSG Malaysia is a wholly owned subsidiary of Nansin Company of Japan. The Japanese company and three Malaysian companies are classified as large-scale businesses, whereas the nine locally owned firms are small/medium enterprises in terms of paid-up capital and number of workers. Fung Keong is the largest local company in terms of paid-up capital with an investment of RM20 million but it employs only 60 workers, compared to RM30 million paid-up capital and 190 employees of the Japanese-owned company. The largest employer by far is Everthrough Rubber Products, a manufacturer of inner tubes that has 550 workers on the payroll and paid-up capital of RM10 million.

There is a domestic motorcycle manufacturing industry that provides a market for original equipment tyres and there is a substantial market for replacement tyres within Malaysia. Three local businesses sell all of their factory output into the domestic market. In terms of exports, the inner tube maker, Everthrough Rubber Products sells 60 per cent of its production overseas, and three companies export between 20 and 40 per cent of production. The Japanese firm NSG Malaysia makes castor wheels that it exports solely to Japan. Malaysian exporters, on the other hand, export to the USA, European Union, Australia/New Zealand, ASEAN countries and the rest of the world, including countries in the Middle East and Africa as well as Pakistan and Brazil.

Latex Dipped Goods Sector

There are a total of 111 companies in the dipped goods sector. Table 8.2 shows the range of products made by the dipped latex line process. Some companies are manufacturers of a range of dipped goods, whereas others are business enterprises making only one product. Latex gloves are used extensively in the medical, dental and healthcare sectors, and in manufacturing industries requiring a sterile or hygienic environment, such as food preparation and the manufacture and assembly of electronic equipment.

Table 8.2 Latex Dipped Goods

Products	*No. of companies*
Medical examination and surgical gloves	60
Household gloves	23
Industrial gloves	29
Clean room gloves, finger cots	21
Catheters and surgical tubes	7
Condoms	15
Feeding teats and babies' dummies	3
Toy balloons	6
Swimming caps and pool socks	3

The breakdown in terms of ownership is 88 Malaysian-owned firms and 23 enterprises with foreign capital investment, either as wholly owned subsidiary companies or in joint venture with Malaysian partners.

Table 8.3 shows the wide range of dipped products manufactured by joint venture and foreign-owned companies. Investment from overseas sources is more or less evenly distributed among the United States (six firms), the European Union (seven firms) and Japan/South Korea (six firms) with two companies with investment from Australia/New Zealand and another two businesses owned by international private investors, including Malaysian interests.

Table 8.3 Dipped Goods Sector: Foreign and Joint Venture Companies

Company	*Source of of investment*	*Products*
Ansell	Australia	Medical, household and industrial gloves
Skellerup	New Zealand	Toy balloons, swimming caps, pool socks
Bard	USA	Catheters, urological tray kits
Bonric	USA	Medical and clean room gloves
Cardinal Health	USA	Medical and clean room gloves
Flexitech	USA	Medical and clean room gloves
Interworld Technology	USA	Clean room gloves, finger cots
Rusch Manufacturing	USA	Catheters, endotracheal tubes
Everts	Germany	Toy balloons
Medical-Latex	Germany	Condoms
Richter Rubber	Germany	Condoms
Mapa Gloves	France	Household and industrial gloves
Marigold Industrial	France	Household and industrial gloves
Unomedical	Denmark	Catheters, endotracheal tubes, tracheostomy tubes
Regent Hospital Products	Sweden	Medical gloves
Dongkuk Techno Rubber	South Korea	Medical and food industry gloves
Sagami Manufacturers	Japan	Condoms
Sanchem Corporation	Japan	Medical and industrial gloves
Shorubber	Japan	Industrial gloves
Sumirubber	Japan	Household and industrial gloves
Suzuki Latex	Japan	Clean room gloves, finger cots
WRP Asia Pacific	USA/UK	Medical and industrial gloves
WRP Sinetimed	USA/UK	Clean room gloves, finger cots

In contrast, Malaysian-owned businesses are mainly concerned with the production of examination and surgical gloves for the medical and dental professions, and clean room gloves and finger cots used by workers in industries processing food and the manufacture and assembly of electrical and electronic products. Table 8.4 shows the range of latex products manufactured by local companies.

A feature of the dipped goods sector is that there are five latex glove manufacturers that are quoted on the Malaysian stock exchange. These companies are large in size in terms of their capitalization and the number of workers they employ compared to the rest of the industry. The largest firm is Top Glove that employs some 6,500 workers and has seven factories in Malaysia as well as manufacturing facilities in Thailand and southern China. Other publicly listed companies are Supermax with six factories, APL Industries with three, Seal Polymer with two and Rubberex with a single production unit.

Table 8.4 Dipped Goods Sector: Malaysian Companies

No. of companies	*Products*
55	Examination, surgical and clean room gloves
11	Household and industrial gloves
11	Condoms
3	Catheters
3	Toy balloons
2	Baby teats
2	Swimming caps, pool socks
1	Medical breathing bags

The five companies with a stock market quotation have greater amounts of capital investment than foreign and joint venture enterprises. On the other hand, when a comparison is made between Malaysian companies that are not listed on the stock market and foreign and joint venture firms, then the position is reversed. In this case, companies with investment from overseas sources are more heavily capitalized than local businesses. In terms of the number of employees, there are substantial differences among the three classes of company. Firms quoted on the Kuala Lumpur stock market typically employ more workers than joint venture and foreign-owned businesses which, in turn, generally have a larger workforce than privately owned local firms.

There appears to be no difference in the amount of factory production sold into overseas markets between the two categories of business. Two-thirds of Malaysian companies and over 80 per cent of firms with foreign capital provide figures on the percentage of production as export sales. All companies export more than half of their output and the great majority of cases (95 per cent) export between 75 and 100 per cent of production.

The conclusion that the dipped goods sector is heavily export-oriented is borne out by the data on the export markets recorded by individual companies. Out of a total of 88 Malaysian enterprises, 82 of them (93 per cent) and all 23 companies with overseas involvement report that they sell their products into markets across the world. The largest market for both local and joint venture and foreign businesses is the European Union, followed in importance by USA/Canada and the rest of the world. For companies with overseas involvement, the Japanese and South Korean market is more significant than it is to local companies. The figures are probably a reflection of the fact that five enterprises are offshore subsidiaries of Japanese companies that sell into the Far Eastern market. The ASEAN regional market also is of greater importance to joint venture and foreign companies than it is to local firms. However, Thailand and Indonesia are major producers and exporters of dipped goods in their own right and three large Malaysian companies operate dipped goods factories in other ASEAN countries.

Latex Other Products Sector

The 16 firms in this sector are divided into 13 locally owned businesses and 3 companies that are either wholly owned subsidiaries of foreign firms or joint venture enterprises. The breakdown in terms of products and ownership is given in Table 8.5. A small number of producers of dipped goods supplying the hospital and healthcare market also manufacture medical items such as latex sheets and dental dams.

Table 8.5 Latex Other Products Sector: Products and Ownership

Products	*Malaysian companies*	*Foreign/ joint venture*	*Source of foreign investment*
Extruded latex thread	4	0	
Foam mattresses and upholstery	4	1	Netherlands
Mattresses and car seats made from rubberized coconut coir	3	0	
Latex rubber sheets, dental dams, medical straps, exercise bands	2	1	USA
Latex toys	0	1	Germany

The largest firms are those that manufacture extruded latex thread used in industries making elasticated textiles for garments such as underclothes, elasticated surgical and sports supports, and elastic netting for wrapping fresh and cooked meat products. Details of the four manufacturers of latex thread are given in Table 8.6.

Another large manufacturer is the American HCM-Hygenic Corporation, a producer of dental dams and tourniquet straps that employs 335 workers although the company does not disclose its capital investment. The Malaysian-owned company, Aerofoam that makes foam mattresses, pillows, cushions and upholstery is a substantial business, with

Table 8.6 Extruded Latex Thread Companies

Company	*Year established*	*Paid-up capital (RM million)*	*Number of employees*
Filati Latex	1988	48	90
Heveafil	1973	84	550
Rubber Thread	1990	14	140
Rubberflex	1986	22	650

RM10.5 million paid-up investment and 160 employees. However, the majority of the local enterprises are small and medium in scale.

Three elastic thread companies, Filati Latex, Heveafil and Rubberflex record that they export basically all (95–100 per cent) of their factory output. The other company, Rubber Thread, does not provide figures on the volume of production that is exported. However, all the manufacturers report that they export thread on a worldwide scale. Rubberflex, for example, exports to over 100 countries, and Heveafil sells into North and South America, Europe, African and Asian countries, the Middle East, Australia and New Zealand. The fully US-controlled firm, HCM-Hygenic Corporation has no directory entry for exports as a volume of factory production but the company exports to Australia, Belgium, Germany, Japan, South Korea and the USA.

Low Technology General Goods Sector

There are 38 Malaysian companies and 9 firms with overseas investment that produce a range of low-tech products, from pencil erasers to door stops and protective tips for furniture. Other items made by this sector include rubber bands and straps, audiovisual equipment surrounds, carpet underlay, flooring tiles, safety tiles for sports arenas and playgrounds, mats for use in bathrooms, kitchens and automobiles, swimming accessories such as swim fins and goggles, toys for children and dogs, sports balls, sealants and adhesive products. Other products made by the low-tech sector include soling sheets for footwear manufacture, sole and heel units, solid tyres and compounding materials.

Out of the nine joint venture and foreign-owned businesses, seven have investment from the Far East, including five firms with share capital from Japan, and one each from South Korea and Taiwan. The other two companies are offshore operations of German companies. A subsidiary company of the Japanese tyre multinational, Bridgestone Sporting Goods produces golf balls for export to the home market in Japan. Another Japanese firm is Central Elastic Corporation, a major exporter of rubber bands and packaging straps to Europe, North America and Australia as well as Japan. Gaskets and stoppers for medical syringes and pharmaceutical containers are made by the Korean company, and the Taiwanese firm produces loudspeaker surrounds for audiovisual systems. The pencil manufacturer, Faber-Castell of Germany concentrates its production of pencil erasers in Malaysia where it has a factory employing 450 workers, which is the largest facility of its kind in the world. Adhesive tapes are produced by the German company, Tesa Tape and three local firms, Central Industrial Corporation, Loytape Industries and Sweet Tape Enterprise.

The largest Malaysian business in terms of capital is LBR Industries, a manufacturer of floor tiles, and car and door mats, with RM22.5 million paid-up capital, whereas the largest local employer is Loytape with 250 workers.

Seven local and five joint venture and foreign companies record that they export from 75 to 100 per cent of production. Details of export sales are provided by 24 Malaysian companies and 7 firms with an overseas interest. Malaysian exporters are represented in all export markets from North America through Europe, the Middle East, Pakistan to China, Japan and South Korea, in the Far East. The most important markets for foreign and joint venture enterprises are Japan and South Korea, which presumably is a reflection of investment by these countries in the sector. This class of business enterprise also has a strong presence in Third World and ASEAN markets.

Footwear Sector

There are 19 companies manufacturing a range of footwear, from rubber-soled sports shoes and trainers to specialist safety boots, as well as companies making sole and heel components for manufacturers and repairers of boots and shoes. Two companies have investment by overseas capital: Bata founded in 1937, the long established manufacturer of shoes, sandals and slippers, and Harvik, a safety boot manufacturer, incorporated in 1973. Bata was originally a Czech-owned business but now has its international headquarters in Toronto, Canada, whereas Harvik was established as a joint venture between the former British plantation group, Harrisons & Crosfield, and Norwegian interests. Of the 17 Malaysian companies 5 firms began operations between 1970 and 1984; and a further 7 businesses were established after 1985 when the Industrial Master Plan came into force. The military boot manufacturer, Kulitkraf was founded in 1968 to supply boots to the Malaysian armed forces.

An analysis of the footwear sector is hindered by the scant information on business activities provided by a number of Malaysian companies in the directory. The conclusion to be inferred from the lack of published data is that these locally owned firms are small in size and they produce footwear mainly for domestic sales in Malaysia.

Malaysian manufacturing enterprises employ fewer workers than the two firms with foreign involvement. The mean number of workers in 12 Malaysian companies is 92, compared to 300 workers employed by Harvik and 510 by Bata. Table 8.7 compares the amount of capitalization and size of workforce of the two large-scale, locally owned companies, Kulitkraf, and International Footwear with Bata and Harvik.

Table 8.7 Footwear Sector: Large Firms

Company	*Source of foreign investment*	*Year established*	*Paid-up capital (RM million)*	*Number of employees*
Bata	Canada	1937	1.5	510
Harvik	Norway	1973	8.0	300
International Footwear		1971	3.0	168
Kulitkraf		1968	7.2	170

The directory entries on export markets show that Harvik, the Malaysian-Norwegian joint venture is an export-oriented company that sells 97 per cent of its industrial safety boots into markets in all four corners of the world. Bata is unique in that not only is the company a manufacturer of footwear but it owns and operates consumer retail outlets under the Bata brand in Malaysia and Singapore. The company has over 200 shops in small Malaysian towns as well as retail outlets in shopping malls in Malaysia and Singapore selling shoes and sandals directly to the general public. Eighty per cent of production is sold in the domestic market and the remaining 20 per cent across the Causeway in the Republic of Singapore. Kulitkraf and Sepatu Timur, two manufacturers of boots for police and military forces, export 10 per cent and 5 per cent of their production respectively into ASEAN, Middle Eastern and Pacific Island countries.

Ten local companies are exporters of footwear, or sole and heel components. The most important markets for made-in-Malaysia shoes, sandals and other footwear are those countries classified as "the rest of the world" including China, Hong Kong and Taiwan, countries in the Middle East and island states in the Pacific Ocean; other major markets are nearby ASEAN countries, Australia and New Zealand.

Compounding and Retreading Materials Sector

There are altogether 21 firms in the compound/retread materials manufacturing sector, with 20 wholly owned by Malaysians and 1 joint venture with Japanese investors. There are 11 businesses which produce both retreading materials (precured treads, sidewall veneers and cushion gum) for sale to the tyre retreading industry, and compounds and masterbatches used as factory inputs by other manufacturing companies. Seven firms are specialist compounders that restrict production to carbon black masterbatch, other masterbatches and a range of compounding materials. Three firms make products solely for tyre retreaders. Other products made by a small number of companies in the sector include solid tyres, inner tubes, tyre flaps, sports balls and moulded rubber items.

The compounder, Lam Seng Tokyo Zairyo Zeon, established in 1994, is a medium-sized, joint venture company with RM2.45 million paid-up capital and 150 employees. It was formed by a local rubber processing company in partnership with three shareholders from Japan. Lam Seng supplies natural rubber from its SMR plant and the Japanese companies provide synthetic rubbers and compounding ingredients for the manufacture of masterbatches and compounds. The company concentrates on the production of compounding inputs for local manufacturers of tyres and tubes, and domestic firms making automotive, engineering and industrial rubber products, such as conveyor belts, hoses, weatherstrip and engine mountings.

Four of the Malaysian businesses are long established, being founded before 1970, including Nam Bee Rubber Works established in 1938. The smallest firm, TKSS Masterbatch, a producer of custom-made coloured compounds, employs only 12 workers, compared to 400 in Goodway Rubber Industries, a major manufacturer and exporter of tyre retread materials. Half of the wholly Malaysian-owned companies are classified as large in size and half are small and medium-scale enterprises.

When the figures for overseas sales as a percentage of factory output are examined, 15 enterprises undertake export sales whereas 5 firms provide no information on their sales to foreign buyers. Two local manufacturers of retreading products, Goodway Rubber Industries and Tread

Rubber Industries export 75 and 90 per cent of production respectively. Three firms export between 50 and 74 per cent of output, and five companies sell between a quarter and half of their production into foreign markets. All 15 firms sell into the market designated as "the rest of the world". Other major markets are South Korea and Japan, Australia and New Zealand, and ASEAN, with the North American and European markets being of lesser importance.

Conclusion

The comparison made in this chapter between foreign and joint venture companies with firms owned by Malaysian interests indicates that generally business enterprises with overseas investment are more heavily capitalized and employ more workers than locally owned businesses. In the special case of the small number of dipped goods manufacturers quoted on the stock exchange, these publicly listed Malaysian firms are larger in size in terms of paid-up capital and number of employees than joint ventures and foreign-owned companies. Both Malaysian companies and firms with foreign involvement are strongly export-oriented and it is not possible to discern any overall difference in export sales policies even though there are differences in each of the seven sectors.

CHAPTER 9

Structural Dualism and Technology Transfer

Introduction

The discussion in Chapters 6, 7 and 8 has considered industrial structure and export sales strategy in 340 manufacturing companies and demonstrated that the industry is highly differentiated by sector. This chapter summarizes the results obtained in the previous three chapters in order that conclusions may be drawn for the entire rubber manufacturing industry.

Two specific topics are investigated relating to the structure of the industry. The first question examines the rubber manufacturing industry as a whole to ascertain if there is industrial dualism between local companies and those with foreign capital investment on the basis of company size and sales to export markets. The second area relates to the automotive, engineering and industrial components sector. It is a more detailed and in-depth investigation exploring this sector's sources of technology and its linkage effects with other sectors of the economy.

Capitalization and Workforce

The section considers whether there is a difference in company size, measured by the degree of capitalization and number of employees, between wholly Malaysian-owned firms and foreign-owned and joint venture enterprises in order to draw definitive findings about dualism in the rubber manufacturing industry.

In statistical calculations a limited number of special cases with figures that are much greater or less than the majority of the cases in a population are termed "outlier cases". Outliers are normally excluded from certain statistical analyses because to include them distorts the results which can lead to an erroneous conclusion being reached. In the automobile passenger car and truck tyre industry there are only three tyre makers and these firms have a significantly larger capital investment and workforce compared to the other manufacturing sectors. Similarly, there are a small number of enterprises producing latex products that are much more heavily capitalized and employ larger numbers of workers than the majority of other firms. In the industrial components sector, one Malaysian company, Kossan Rubber Industries has much greater paid-up capital than the rest of the sector. All these businesses have therefore been excluded from the analysis of capitalization and size of workforce that is given in Table 9.1.

There are 238 companies that publish details of paid-up capital after excluding outlier cases. Businesses totalling 181 are controlled by local interests and another 57 have investment by foreign capital. Statistical analysis demonstrates that there is a significant difference between paid-up capital for Malaysian (mean = RM2.9 million) and joint venture and wholly foreign-owned companies (mean = RM10.6 million). The conclusion is drawn that joint venture and foreign enterprises have a greater capital investment compared to business concerns owned by Malaysian investors.

Table 9.1 Rubber Manufacturing Industry: Capitalization and Number of Employees

	Malaysian		*Foreign/Joint venture*	
	mean	*n*	*mean*	*n*
Capitalization (RM million) 238 companies	2.9	181	10.6	57
Number of employees 281 companies	158	217	418	64

Note: Excludes outlier companies.

There are 281 companies, excluding the outlier manufacturers, that provide information on the number of employees. Analysis of these companies shows that foreign and joint venture companies employ a statistically greater significant number of workers than local firms. There are 64 businesses with foreign capital employing a mean figure of 418 workers, compared to 217 enterprises controlled by domestic interests that have a mean labour force of 158 employees.

Out of the total survey population of 340 companies, 309 firms, including outlier companies, provided information on paid-up capital and size of the workforce. These companies have been placed into two categories of large-scale enterprises and small/medium-scale enterprises as shown in Table 9.2. The figures demonstrate that 79 per cent of foreign companies and joint ventures are classed as large, compared to 34 per cent of Malaysian firms. Statistical analysis indicates a significant difference in size between the two classes of company. The conclusion is that the proportion of large-scale foreign and joint venture companies is significantly higher than the proportion of large-scale Malaysian companies.

Table 9.2 Large and Small/Medium-scale Enterprises: Sector and Ownership

	Malaysian		*Foreign/Joint venture*	
	Large	*SME*	*Large*	*SME*
Tyre industry				
Pneumatic tyres	1	0	2	0
Motor cycle/solid/tubes	3	9	1	0
Latex industry				
Dipped products	45	39	19	3
Other products	5	8	2	0
General products industry				
Industrial products	12	55	24	7
Low tech products	5	27	5	4
Footwear	2	10	2	0
Compounding industry	9	9	0	1
Total	82	157	55	15
Per cent	34%	66%	79%	21%

An axiom in the study of industrial economics is that in the production process the firm uses a balance of capital and labour, depending on the relative cost of each factor, in a bid to maximize profits or minimize costs. Thus a manufacturing enterprise may adopt a high technology, capital-intensive mode of production and employ relatively few workers. Alternatively, a manufacturer may employ a large workforce and rely on less capital-intensive factory operations in situations where the cost of wage labour is relatively cheap compared to capital. It is possible to use the mean figures for capital and labour to calculate the amount of capital employed per worker as an approximation of capital/labour ratio for the rubber manufacturing industry including statistical outlier companies. The mean figures for Malaysian businesses are RM8.1 million paid-up capital and 211 workers. For foreign and joint venture companies the figures are RM20.5 million and 462 employees. Calculation of the capital/labour ratio shows that for local firms each unit of labour employs RM38,400 paid-up capital, whereas for firms with overseas investment the ratio is one worker per RM44,400 capital investment. The conclusion is that although joint venture and foreign-owned firms employ more workers than Malaysian companies they invest a greater amount of capital per worker than local firms. These figures, therefore, suggest that the labour productivity of foreign and joint venture enterprises is higher compared to Malaysian-owned manufacturing concerns. Malaysian companies, on the other hand, are undertaking factory operations that are inherently more labour-intensive than manufacturers with foreign investment.

The forgoing analysis has confirmed the assertions made by UNIDO and the MRB that there is a dual structure in the rubber manufacturing sector, with firms employing overseas capital being larger in size than Malaysian companies. Foreign and joint venture companies are more heavily capitalized and employ a greater number of workers than wholly Malaysian-owned business enterprises. They are also more productive because they employ more capital per worker than local manufacturers.

Management

Another factor that influences productivity is the management of a company and how executive managers use its assets of capital, labour and technology. There are clear differences in the standard of management between international manufacturers and larger Malaysian industrial enterprises on the one hand and smaller poorly capitalized local firms on the other. On the whole, larger firms, both foreign and Malaysian, tend to employ more highly educated personnel compared to the smaller local business enterprises.

In foreign-owned firms top management is typically provided by international businessmen employed on generous expatriate terms of service such as free housing, paid passage for families for home leave and assistance with school fees. Such personnel almost invariably are educated to tertiary level with many having postgraduate qualifications in business administration. The posting to Malaysia for the majority would be regarded as being a step-up on the career ladder within a transnational company that operates in a number of countries worldwide. Examination of directory entries show that Japanese manufacturers, in particular, employ Japanese nationals in a number of managerial positions, whereas other foreign firms tend to have fewer expatriate personnel apart from key positions such as chief executive officer and specialist technical managers. International companies normally recruit their Malaysian executive staff with university degrees and typically provide more generous terms of employment than local firms.

Large local manufacturers, including family-owned businesses, employ Malaysian nationals in managerial and technical positions who generally are educated to the same level as their counterparts in international companies. Many of the local managers are graduates of overseas universities with qualifications in technical subjects and business studies. The Malaysian university system also provides degree level courses in business and technology. By way of contrast, managers in small family-run businesses, typically, have been educated to secondary level in Chinese-medium schools although a number are graduates of universities in Taiwan.

Export Sales

The most important export sector is the latex products industry that accounts for 75 per cent by value of total exports. Companies that manufacture general rubber products account for 17 per cent of the value of exports; the tyre and inner tube sector accounts for 7 per cent of export revenues. Data on the volume of factory production sold into export markets are provided by 178 manufacturers in all product sectors. The breakdown is 129 local concerns and 49 joint venture and foreign companies. When a comparison is made between the two categories of firms, a clear difference in business strategy is observed. In the case of companies employing foreign capital, 90 per cent of firms export 50 per cent or more of production, compared to 70 per cent of businesses owned by Malaysians, as shown in Table 9.3. The conclusion is that foreign and joint venture enterprises export a greater volume of factory output than local firms, even though more than two-thirds of Malaysian firms export more than half of total production. In fact, for Malaysian manufacturers of dipped goods and extruded thread they export basically all of their production, as do foreign-owned and local producers of latex gloves and medical goods.

Table 9.3 Exports as Percentage of Production

Exports	*Malaysian*	*Foreign/Joint venture*
Over 50%	91 (70%)	44 (90%)
Less than 50%	38 (30%)	5 (10%)
Total	129	49

There are 208 Malaysian and 65 joint venture and foreign companies that provide data on the markets in which they make export sales. The percentage of the companies that sell into six market segments is shown in Table 9.4.

The figures indicate that for all markets, except the Japanese and South Korean market, there is little difference between the two groups. The greater proportion of foreign and joint venture enterprises (75 per cent) that sell into the Far Eastern industrialized market, compared to 57 per cent of Malaysian exporters, is a reflection of the propensity of Japanese-owned companies to produce for their home market. Overall, the statistics show that both local firms and those with foreign investment are strongly export-oriented and are not reliant on any one market.

Table 9.4 Export Markets: Number and Percentage of Exporting Companies

Ownership	*Firms*	*USA/ Canada*	*European Union*	*Japan/ South Korea*	*Australia/ New Zealand*	*ASEAN*	*Rest of the world*
Malaysian	208	128 (62%)	154 (75%)	118 (57%)	115 (55%)	133 (64%)	164 (79%)
Foreign/joint venture	65	35 (54%)	39 (60%)	49 (75%)	31 (48%)	45 (70%)	48 (74%)

Made-in-Malaysia products are sold across the world. An unexpected finding of this study is the importance of South-South trade to rubber manufacturers in Malaysia. Over 70 per cent of each category of company report export sales into emerging markets and less developed countries in other parts of Asia, the Middle East, Latin America and sub-Saharan Africa.

Linkage Effects

One of the objectives of the study is to examine the linkages the industrial products sector has with other sectors of the Malaysian economy. Forward linkage effects are measured by an examination of the figures of the volume of factory production sold into downstream industries located in Malaysia. Backward linkages are observed by examining the source of manufacturing inputs, such as natural and synthetic rubbers, compounding materials, and factory machinery and equipment.

In the industrial components sector, there is a clear difference in the volume of production sold into the local industrial market by Malaysian, and foreign and joint venture manufacturing firms. Analysis of sales figures obtained from the trade directories and the questionnaire indicates that over 60 per cent of local companies sell in excess of 50 per cent of their production to other Malaysian-based enterprises. On the other hand, the directories give a figure of only 10 per cent and the questionnaire 25 per cent of firms with foreign involvement that sell more than half of output into Malaysian downstream industries. Table 9.5 has the details. The conclusion may be drawn that Malaysian manufacturers have stronger forward linkages with the local economy than firms with overseas financial interests.

Table 9.5 Domestic Sales: Companies Selling over 50 per cent of Output Locally

Ownership	*Secondary data (%)*	*Primary data (%)*
Malaysian	62	65
Foreign/joint venture	10	25

The questionnaire results indicate that there is no difference in procurement policy for natural rubber between the two categories of company by ownership. Out of the 25 companies that took part in the survey, 14 firms use only rubber produced in Malaysia. Of the remaining 11 firms, 9 use more Malaysian rubber in manufacturing operations than imported rubber. Only two manufacturers use more imported rubber than locally produced rubber. The 11 manufacturers that source *Hevea* rubber from outside the country obtain most of their supplies from rubber producers in ASEAN, in particular Thailand. Two businesses buy rubber from non-ASEAN sources although the producing countries are not identified. Since the largest international market for natural rubber is in Singapore it is relatively easy for Malaysian-based companies to purchase rubber from countries other than Malaysia.

Although Malaysia has a substantial manufacturing sector, the country does not have a well-developed, high technology engineering and machine tool industry. However, given the rate of industrialization of the country there is growing ability to produce capital machinery for the expanding manufacturing sector. Nevertheless, capital equipment for manufacturing operations in most industries is procured largely from the industrialized world, in particular Japan, USA and the European Union. This is the case with the industrial rubber products sector. All companies, whether locally controlled or foreign-owned, purchase over three-quarters, and in many instances 100 per cent, of their factory machinery and equipment from overseas manufacturers. The industrial components production sector, in common with the other manufacturing sectors, does not have strong linkages with Malaysian manufacturers of factory machinery.

When the questionnaire survey was carried out, solid synthetic rubbers were not made in Malaysia, nor were synthetic materials manufactured in Singapore, a major petroleum refining centre, so that supplies of synthetic elastomers had to be shipped in from producers outside the region. At the time of the survey, there were a number of importers of synthetic rubbers that held stock and sold into the rubber manufacturing industries. The questionnaire results show that some 80 per cent of Malaysian rubber manufacturers buy 75 per cent or more of their requirements for synthetic rubbers from local suppliers. This is in contrast to 75 per cent of foreign and joint venture enterprises that purchase most of their synthetic inputs direct from suppliers in overseas countries. The conclusion is that, in the case of purchases of synthetic rubbers, Malaysian companies are more firmly linked to local suppliers than companies with foreign investments.

The study has identified a number of firms making a range of masterbatches and other compounding materials that sell their products to other manufacturers, especially small and medium-sized companies. In addition, there are a number of specialist suppliers of imported processing agents based in the country, and there is domestic production of a limited range of chemicals, including carbon black. There is, therefore, an established network of Malaysian-based firms that supply locally produced and imported compounding inputs to the rubber manufacturing industry.

The questionnaire looked at two aspects of procurement policy of compounding material: (a) are the products manufactured in Malaysia or overseas, and (b) are the suppliers of these materials based in the country or outside it?

On the issue of the source of manufacture, the survey results indicate a difference in procurement policy between the two categories of manufacturer. The majority of Malaysian-owned firms use mainly locally produced compounds and compounding agents in their manufacturing operations. On the other hand, most joint venture and foreign-owned companies utilize more compounding materials made in foreign countries. A similar picture emerges when the location of suppliers of these products is examined. Malaysian-owned manufacturers buy a greater volume of their inputs from local suppliers, whereas companies with overseas involvement purchase their compounding products from suppliers based outside the country. The backward linkage effects for compounding materials are, therefore, greater for Malaysian firms compared to those with financial interests from overseas.

The linkage effects of the industrial products manufacturing sector may be summarized as follows. In terms of forward links, there are strong linkage effects between Malaysian producers of industrial components and local manufacturing industries, in particular the automobile and motorcycle manufacturing and assembly sectors, and exporters of electronic and electrical equipment. By way of contrast, as foreign-owned and joint venture enterprises concentrate more on sales to overseas markets, they tend to have weaker links with other manufacturers in the Malaysian economy. In backward linkages there are both similarities and differences between the two categories of company by ownership. All manufacturing companies have strong backward links with the local natural rubber production industry that supplies the bulk of raw material to the industrial sector. On the other hand, all firms have only weak linkages with Malaysian manufacturers of factory machinery and equipment, with almost all of the required machinery being made overseas and imported into Malaysia. There are differences in linkage effects between the two classes of company with regard to synthetic elastomers and compound materials. Malaysian firms have strong backward linkages with domestic producers and suppliers of compounds and compounding ingredients as well as locally based suppliers of imported synthetic rubbers and specialist chemicals. Joint venture and foreign-owned businesses, on the other hand, exhibit only weak linkages with Malaysian-based companies that supply synthetic rubbers and compounding inputs, and local producers of compound materials.

Technology Transfer

The development programme in the first UNIDO industrial master plan recommended strengthening the industrial capability of research organizations servicing the natural rubber industry. There was to be a concentration on R&D in rubber technology and emphasis on providing technical assistance to manufacturers based in Malaysia. To examine the impact of the government-controlled MRB research institutes on the local rubber manufacturing industry, the postal survey included questions on the transfer of technology and provision of technical assistance.

The MRB has been successful in developing strong technical links with domestic manufacturers for the supply of industrial technology and the provision of technical consultancy services. This is evident in the survey results which show that almost all Malaysian-owned firms and a Singapore-controlled joint venture use the MRB to source their manufacturing technology and obtain technical advice on problem-solving in factory operations. The reason that Malaysian companies have heavy reliance on the MRB for manufacturing technology and technical assistance may be partly a reflection of their smaller size and insufficient scale to fund in-house R&D facilities. On the other hand, the majority of joint venture and foreign-owned enterprises obtain factory technology and technical assistance from overseas partners or parent companies. They have little contact with the Board for either basic factory technology or technical advice. Other important sources of industrial know-how for both classes of company are techniques developed internally in company-owned laboratories and on the factory floor.

The MRB consultancy company, the main source of advice and assistance to local companies, has access to know-how and backup services provided by the Board's research laboratories in Malaysia and England. The two rubber research centres strengthen the technical base of the industry as a whole because their services are open to all manufacturers whether foreign or Malaysian. The access that foreign-controlled firms have to R&D laboratories in their home countries provides another source of technology transfer.

General Conclusions

The chapter has brought together the results of the examination of the business behaviour of all production sectors, and amalgamated the data into an analysis for the entire rubber manufacturing industry. In the literature there is a limited number of general assertions, first recorded by UNIDO and repeated by other commentators, that there is a dual structure to the industry (UNIDO 1991: 75; Taylor & Ward 1994a: 153–5; Tham & Mahani 1999: 65; Abdul Hamid Sawal 2001: 110). These authorities state that there are a small number of large firms controlled by transnational companies and a large number of small and medium-sized enterprises owned by Malaysian business interests. Foreign-owned firms and joint ventures employ more advanced technology than local companies and are stronger in export markets. However, there are no data or other hard evidence presented in the literature to support these assertions that are brief and of a general nature. One of the objectives of the present study is to test these statements by analysing data published for the first time in the MRB directory and by a questionnaire survey.

The results are that the number of foreign and joint venture sectors cannot be described as small as it accounts for 20 per cent or 73 firms out of a total of 340 that makes up the entire rubber manufacturing industry. However, foreign investment is not evenly distributed among the different production sectors. It is concentrated in the automobile tyre and the industrial components industries where there is direct foreign investment in 30 per cent of the total number of companies. The latex dipped goods, latex other products and low technology general product sectors account for 20 per cent of firms with foreign involvement. There is low foreign participation in the motorcycle/inner tube, compound/retreading materials and footwear sectors.

When the industry is taken as a whole, the results of statistical analysis to compare the degree of capitalization and number of employees in each category of ownership show conclusively that joint venture and foreign firms are more heavily capitalized and employ more workers than local companies. When companies are classified into large-scale enterprises and small/medium-sized enterprises, the statistical calculations demonstrate that firms with foreign capital are larger in size than Malaysian businesses. The results are that 79 per cent of foreign and joint venture companies are classified as large whereas 34 per cent of local enterprises are in the large category.

There are differences in the importance of export sales policy between the two ownership groups even though the industry as a whole is strongly export-oriented. Apart from the latex industry, foreign and joint venture businesses export a greater proportion of production than Malaysian companies. In the latex products sector there is no difference in the volume of export sales between the two classes of company. The industry is not dependent on any single market and firms sell their products across the globe, although the American and European markets are the most important in terms of export revenue.

The study, therefore, confirms the conclusion made in the report produced by UNIDO that the rubber manufacturing industry has a dualistic structure and that there are differences in business behaviour based on ownership of assets. Local Malaysian investors mainly control the industry but the foreign and joint venture sector is substantial, accounting for one-fifth of all manufacturing companies. There are clear differences in structure between the two groups of company based on ownership. Joint venture and wholly foreign-owned firms are typically larger in size and export a greater proportion of production than Malaysian companies. Locally owned businesses typically are classified as small/medium-sized enterprises and have stronger linkages with other sectors of the domestic economy compared to firms with overseas investment.

CHAPTER 10

Policy Issues in Industrialization

Objectives of the Study

This chapter brings the results of the analysis together in order to discuss the findings in the context of current theories in development economics and to draw a general conclusion on the role the rubber products manufacturing sector has played in Malaysia's development from the colonial period onwards. The study examines the rubber manufacturing industry with the primary aim of assessing the contribution the sector has made to the industrialization and transformation of the country from a primary commodity producer to a newly industrialized economy. A secondary goal is to investigate whether there is a dualistic structure in the industry by comparing differences in business behaviour between companies with overseas investment and those wholly owned by Malaysian firms. By adopting an overall approach from the standpoint of economic history, the development of the rubber manufacturing sector is analyzed, from the establishment of *Hevea* plantation agriculture at the beginning of the 20th century to the present day. Information published in the MRB trade directory has been used as the principal data base to conduct a micro-economic study of the rubber products manufacturing industry in 2005 when the Second Industrial Master Plan came to a close.

Political Economy of Industrial Policy

A widely held viewpoint of academics working in the field of development studies is that the political economy of a country plays a critical role in the development process. A country's social, cultural and political factors greatly influence the part that its government plays in promoting policies that may enhance or hinder economic progress. In industrial development theory, there is general consensus that direct intervention by government is important in promoting an active industrial strategy in areas such as:

- formulation of foreign trade and overall industrial policy;
- industrial investment policy; and
- policy towards foreign direct investment and treatment of multi-national investors.

The success of governments in promoting industrial development is dependent on the availability of managerial and administrative skills in the civil service, the nature of the country's political ideology and strong governance based on the rule of law.

The political economy of Malaysia since Independence in 1957 demonstrates the pragmatism of decision-makers in establishing policies to promote industrialization within the framework of a socio-economic development programme designed to advance the position of the Malay community. This pragmatic approach has, it is suggested, been heavily influenced by the fact that historically and in contemporary times the country is an open economy fully integrated in the world trading system. In the past, Malaysia imported capital and consumer goods financed by the export of tin and rubber; today imports are paid for by commodity exports, especially crude petroleum oil and gas, and refined palm oil, and importantly by the overseas sales of manufactured goods, including rubber products.

The pragmatic nature of successive governments is demonstrated by the authorities, on occasion, questioning the conventional economic wisdom of the time and rejecting the advice of academic development theorists and international institutions. Thus at the time of Independence, Malaysia embarked on a period of agricultural expansion in plantation crops and rice production rather than encouraging an urban, industrialization programme protected by high tariff barriers. The orthodox opinion of development economists under the structural school is that the agricultural sector and peasant farmers are an economic "milch cow" to supply unlimited numbers of workers to highly protected infant industries while providing the rapidly growing urban population with cheap food. Foreign exchange revenues from exports of commodities together with taxes raised from the rural population, for example, agricultural land tax, are to be used to develop the industrial sector. The realities of Malaysian politics, on the other hand, encouraged the government to follow a heterodox approach and use export earnings from rubber and tin to invest heavily in rural development projects. These favoured politically influential, rural Malay voters over the predominantly Chinese and Indian populations in the towns.

The Asian financial collapse and recession of 1997/98 is another example when the authorities in Malaysia did not accept the recommendations of international economic institutions. The country rejected the advice of the IMF to devalue and bring in an economic austerity programme but instead imposed stringent currency controls and pegged the *ringgit* at a fixed rate against the US dollar. This controversial approach was strongly criticized at the time but with the benefit of hindsight there is a body of opinion that holds that Malaysia was able to recover more rapidly from the crisis than Indonesia and Thailand that did follow the IMF prescription.

On the other hand, the authorities closely followed the recommendations of the World Bank in 1955 to diversify out of natural rubber into oil palm, and establish a relatively small manufacturing base to supply goods for the domestic market. Similarly in 1985, after the failure of the heavy industrialization policy, the government embraced the advice of UNIDO in the Industrial Master Plan to expand industries using natural resources produced in Malaysia and in which the country has a comparative advantage.

Development Policy Issues

Porter Model of Trade and Industrialization

The seminal trade and industry theory formulated by Porter relies heavily on endogenous growth theory for its theoretical assumptions. Growth theory states that investments in knowledge-based agencies that promote research and development, together with education and training in human resources are the principal factors in stimulating long-term economic growth. The theory is rooted in the neo-classical paradigm and emphasizes the benefit of the exchange of ideas that comes with an open economy integrated into the wider global pattern of trade. A policy implication is that governments can promote growth by undertaking an active role in investments in human capital and providing incentives to agents in the knowledge-producing R&D sector.

Porter's theory classifies production factors into two categories: basic and advanced. A country's basic factors are the inherent advantages of geography and demography that require only modest investment to become productive; advanced factors, on the other hand, are knowledge and skill-based. Basic factors conform to the neo-classical comparative advantage theory of international trade. However, Porter's advanced, knowledge-based factors modify classical comparative advantage theory and take it to a higher level of "competitive" advantage. There is, therefore, a distinction between a nation's basic *comparative* advantage endowed by natural resources and its advanced *competitive* advantage subject to intervention by government policies. The revisionist school of neo-liberal economists argue that governments can take an active macro-economic role by formulating a basic industrial development strategy and building on the strategy by the implementation of micro-economic policies to develop a knowledge-based economy to enhance the country's competitive advantage. The Porter national competitiveness model posits that a concentration of business enterprises producing similar products leads to rigorous competition among companies for market share. This, in turn, results in technical innovation leading to the introduction of improved products and entry into new markets all of which stimulate economic growth and the overall development of the country.

British Malaya in the early years of the 20th century had the attributes of the basic production factors of Porter's model to allow the development of a commodity crop economy that produced and exported natural rubber in a laissez faire free market economic system. The equatorial climate of the land abundant but sparsely populated Malay Peninsula was ideal for the establishment of large-scale rubber plantation agriculture. Capital raised mainly on the London Stock Exchange was invested in plantation companies that were alienated land by the British colonial administration, which also allowed entry of South Indian and Chinese labour to work the estates. Rubber, an essential raw material for the production of motor vehicle tyres, was transported from the deep water ports of Penang and Singapore, situated on the shipping lane between East and West, to tyre manufacturers that supplied the rapidly expanding automobile industries in the USA, Europe and Japan.

At Independence in 1957, Malaysia had a number of Porter's advanced production factors in place to give the country a competitive advantage over other rubber producers in Southeast Asia. The country was the world's largest natural rubber producer that exported rubber through an established international commodity marketing system. Rubber production on plantations was heavily capitalized and employed a professional managerial class of expatriates, a local staff of supervisors, clerks and skilled artisans, and a disciplined, wage-earning labour force. There was a basic infrastructure already in place of roads and railways that linked the hinterland to Penang in the north and Singapore in the south of the Peninsula. The development of the rubber manufacturing industry was enhanced by technological discoveries on the use of rubber as an industrial raw material. In Malaysia, the Rubber Research Institute, established in 1925, had as its remit research into the agricultural botany of rubber cultivation and the technology of crop processing. The UK-based British Rubber Producers' Research Association (now the Tun Abdul Razak Research Centre, TARRC) was founded in 1938 to carry out fundamental scientific research into the properties of natural rubber and applied studies in rubber technology. Natural rubber production and the manufacture of rubber goods in the 1950s therefore had a solid scientific foundation on which to build in more recent years.

The UNIDO development strategy of 1986 recommended in the First Industrial Master Plan drew heavily on the neo-classical economic paradigm, endogenous growth theory and the Porter model of industrial development. This overarching strategy continued to be implemented in the Second and Third Plans, and in the Economic Transformation Programme of 2010. Specific policies for the rubber manufacturing sector include:

- resource-based industrialization using natural rubber in which Malaysia has a competitive advantage;
- export-oriented trade in manufactured goods produced by the rubber-based industries;
- investment in technology and R&D activities to support the rubber manufacturing industrial sector;
- encouragement of direct foreign investment by international rubber products companies to gain access to advanced manufacturing technology and export markets; and
- development of human resources and training for the industry.

Importantly, all programmes adopt the Porter development model for clusters of related industries to stimulate manufacturing companies within the group to compete in developing high-value, technologically advanced products.

Balanced Growth Model

All three industrial master plans recommend the simultaneous development of a range of industries, both resource and non-resource-based, in a policy similar to the balanced growth model of investment in a number of industrial sectors that was current in the 1950s. Natural resource-based industries are regarded as naturally evolving enterprises utilizing

Malaysia's resource availability and comparative advantage. The manufacture of products made from rubber, palm oil and timber are highly dependent on local factors, including raw material inputs, domestic capital investment and local technology. Among the non-resource-based industries are the multinational electronic component and electrical appliance businesses in Free Trade Zones, and the automobile and steel industries created by government policy under the heavy industries programme. These are highly dependent on foreign technology. It may be argued that this multi-industry development strategy adopted by Malaysia builds on the strengths of the different industrial sectors, and each sector may be regarded as a complementary component within the overall politico-industrialization programme for the country to achieve fully developed economic status by 2020.

Foreign Direct Investment

The political ideology followed by governments is crucial in deciding the policies to be adopted towards multinational companies and foreign direct investment. The authorities in Malaysia from colonial rule to the present day have been welcoming the investments of foreign companies as a source of external capital, manufacturing technology and management expertise in the process of industrialization. Immediately after Independence, the government passed the Pioneer Industries Ordinance (1958) that allowed exemption from corporation tax and tariff protection for newly established industries. With the promulgation of the New Economic Policy in 1970, transnational investment in industry was encouraged to counter the economic dominance of the Chinese business sector. The government established Free Trade Zones for light industries where multinational corporations assembled electrical and electronic products using imported components for export. Foreign companies were exempt from custom duties on imported parts, and received fiscal incentives to encourage them to establish production facilities in free trade manufacturing enclaves. Investment in joint ventures by South Korean and Japanese conglomerates in heavy industry was a feature of the "Look East" policy between 1980 and 1985. When the First Industrial Master Plan became operational in 1986, the Promotion of Investment Act was passed to provide incentives for investment by overseas companies. The legislation allows majority foreign control for firms that export at least 50 per cent of factory output, while foreign manufacturers can set up wholly owned subsidiaries if more than 80 per cent of production is sold into export markets. The success of the policies to encourage direct foreign investment can be judged by the fact that 20 per cent of companies in the rubber manufacturing industries are either joint ventures with Malaysian investors or wholly foreign-owned. In the export-oriented industrial rubber products sector that makes high specification, intermediate components for other industries, the figure for foreign participation rises to 29 per cent of all firms.

Globalization

The forces of globalization that have impacted on the world economy in recent decades have also influenced the ownership and structure of the rubber manufacturing industries. The story of Continental Sime, for example, encapsulates within a single enterprise how the process of globalization has occurred in the automobile tyre sector. The company was founded in 1961 as a subsidiary of the British tyre maker, Dunlop to produce car and lorry tyres to supply the domestic market. In 1984, when Dunlop sold its European and American interests to Sumitomo of Japan, the Malaysian factory was acquired by Sime Darby, a diversified plantation, industrial and trading group. Sime Darby already had a wholly owned tyre company, Sime Tyres International, established in 1979, that it continued to operate as a separate enterprise with the then publicly listed Dunlop Malaysian Industries. In 2003, the two firms were amalgamated when Sime Darby entered into a joint venture with the Hanover-based Continental AG, in which Continental owns 51 per cent of the equity. The partnership has enabled the German tyre manufacturer to enter the ASEAN and Australasian markets where it previously had only a limited presence, and Sime Darby has access to advanced tyre making technology of an established European multinational corporation.

There have also been mergers and disposals of companies in the latex goods industry. The London Rubber Company established production facilities to produce its Durex brand of sheath contraceptives, Marigold industrial and household gloves, and Regent surgical gloves in Malaysia. After a series of mergers in the healthcare industry in the UK, the successor company in 2003/04 moved the Durex condom manufacturing operations to southern Thailand and sold off its glove factories to French and Swedish interests that continue to produce latex gloves under the Marigold and Regent trade names. The American medical supply company, Baxter also set up a number of dipped product factories in the 1980s that it subsequently sold to other American manufacturing firms.

In the extruded thread sector, the four companies are wholly Malaysian-owned, although Heveafil and Filati Lastex were initially established as joint venture enterprises between the government-controlled PNB and Mardec corporations with Italian thread manufacturers. All producers, however, rely on Italian technology for extrusion manufacturing operations and use factory machinery imported from Italy.

Policy Implications of Economic Development

The Political Imperative

The policy dilemma facing developing countries with an abundance of mineral and agricultural resources is whether to go down the route of resource-based industrialization or concentrate on the production of primary commodities for export in their endeavour for economic development. This basic decision on trade and industry policy is unique to each and every sovereign nation state, and will differ depending on the level of political and economic development as well as the underlying political ideology of individual countries. Countries have to devise their own strategies for industrial expansion and export-led growth. This normally involves some form of development plan in which policy-makers analyze the problem, decide on priorities and seek to remove any obstacles in the way of achieving success. Once a country has decided where the opportunities for growth are, it has to choose quite deliberately the sectors it wishes to nurture and go about building them. Favourable tax, regulatory and planning policies do not happen by themselves and investment in basic infrastructure and business parks require public finance. Politicians and government officials, therefore, have to have firm commitment to implement such changes and have a clear plan where the country's best long-term interests lie. What is important is that strong political commitment is essential for the implementation of a successful industrialization policy.

For countries deciding to turn their resource base into manufactured goods, the experience of Malaysia over past years may point to some lessons to be learnt. Malaysia is a resource-rich country, endowed with abundant supplies of natural rubber, palm oil, petroleum oil, natural gas and tropical timber. The discussion in earlier chapters has demonstrated that such a wealth of natural resources (including tin in former years) made possible rapid economic growth on the basis of primary production, although it may be argued, weakened the imperative to industrialize, compared to resource-poor countries in Asia, such as South Korea. Exports of primary commodities have been crucial for the financial viability of Malaysia since colonial days. The revenues from these exports have been prudently utilized so that when the decision was made to industrialize there was an existing institutional framework and countrywide infrastructure. Resource rents in the form of export duties imposed by successive governments have been invested in physical infrastructure crucial to capital accumulation: electricity generating plants, industrial estates, ports, roads and railways. Export revenues have also been used to finance education and other essential social services, and build capacity in the civil service for effective governance of the country. Malaysia was thus able to successfully diversify from dependence on trade in commodities such as rubber and palm oil into the more recently developed downstream activities of rubber goods manufacture and oleo-chemical refining, to increase value added.

The active role of government has been crucial in achieving this structural transformation and degree of industrialization. State intervention in industrial policy was especially pronounced from 1970 up to 1985 under the New Economic Policy when the authorities gave priority to inter-ethnic redistribution of wealth in favour of the Malays. In the case of the heavy industrialization programme in the early and mid-1980s, politicians steered the policy, with little input by economists or technocrats. The programme used finance from government sources and co-opted multinational conglomerates from South Korea and Japan to provide technical expertise and manufacturing inputs in new industries, such as motorcar manufacturing and steel plants. However, the internal pricing policies of the foreign companies and intense competition on world markets resulted in a lack of international competitiveness for the products of the new industries. The government was quick to rectify the failures in the heavy industry development strategy by seeking advice and recommendations from outside expertise. It was not until the first UNIDO ten-year Industrial Master Plan of 1986 that industrial policy interventions were conceived and implemented on a more rational basis.

In common with many developing countries, Malaysia has a small domestic market such that the internal demand for manufactured goods is easily satisfied. The Chinese entrepreneurial community, which controls much of the commercial and industrial activity, has only limited technological capability and managerial expertise, and is weak in international marketing networks. Compared to other ASEAN nations, the Malaysian workforce is relatively well paid; hence the country is at a competitive disadvantage against neighbouring low-wage countries. The policy which the Malaysian government has adopted to overcome these constraints is to encourage the establishment of export-oriented industries that manufacture medium to high-value products. The strategy has been to actively attract foreign transnational companies to invest in new manufacturing concerns, particularly in the resource-based sector. The priority accorded to the inflow of foreign investment is based on the assumption that manufacturers from overseas will stimulate technological upgrading of manufacturing operations, introduce modern management practices and encourage export-led growth. This emphasis on the process of resource-based industrialization is to enhance the natural comparative advantage of the existing resource base. The outcome is for well managed, technically advanced factories to produce high value-added manufactured goods that are able to compete successfully on the world market. One other feature of Malaysia's approach to overseas involvement in industry is that the authorities have welcomed investment from transnational firms from varied sources in eastern and western industrialized nations. This policy, it is argued, may have increased the country's leverage and bargaining power in negotiating the entry of foreign capital.

The foregoing discussion argues that Malaysia's natural resource endowments have been used as part of an overall development strategy to diversify the economy away from primary commodity exports into value-added manufactures. The development of resource-based manufacturing industries means that the country will increasingly supply the international market with rubber goods instead of raw rubber, specialist oleochemical products derived from palm oil and wooden articles made from tropical timber in a fully industrialized economy. The policy of encouraging foreign investments in selected industries to increase productivity leads to higher profitability that maximizes gains for the national economy. It is argued that the country has progressed well beyond Porter's basic comparative advantages derived from natural resource endowments to enjoy competitive advantages with the development of industrial capacity in a range of manufacturing industries in an advanced economic system.

Policy towards Foreign Investment

The Malaysian authorities in past years and during present times have been welcoming to overseas investment from a variety of countries. There has, therefore, been no strong movement against foreign direct investment in manufacturing industry on political or ideological grounds by the ruling establishment in the country. On the contrary, it may be argued that, besides the accompanying expertise they bring, the entry of overseas firms has been encouraged by decision-makers as a countervailing economic force against the dominant ethnic Chinese-owned business sector, thus strengthening the position of the main Malay political party within the complex inter-ethnic political discourse of Malaysia. This situation is not paralleled in many other developing countries where fervent nationalist sentiments dominate the political debate and there may be a reluctance to admit foreign capital because of a fear of neo-colonial interference in domestic affairs. The policy implications for the implementation of industrialization strategies in developing countries, therefore, seem to be clear. Governments wishing to allow transnational enterprises to operate need to decide on an ideological commitment to accept some degree of foreign participation in the economy. They then have to lay down firm ground rules enacted through legislation that govern the entry and economic activities of overseas investment in the host country.

Utilization of International Agencies

The government of Malaysia has utilized the resources of international development agencies to assist in the formulation of policies to promote industrialization. As far back as 1955, the colonial authorities appointed the World Bank to prepare proposals for industrial development and the report became the policy guidelines for the newly independent Federation of Malaya. The industrial strategy recommended the establishment of a limited number of manufacturing industries to satisfy the domestic market for consumer and industrial goods, including motor vehicle tyres. After the failure of the heavy industrialization programme in the 1980s, the authorities commissioned UNIDO to undertake a root and branch review of the country's industrialization policy. The recommendations contained in the first ten-year Industrial Master Plan provided a detailed development strategy for the rubber manufacturing sector that was included in a cluster of seven resource-based industries. The success of the industrialization strategy recommended in the report led to a second plan for the period 1996–2005 in which the rubber manufacturing industries were again identified as a priority sector in the group of natural resource-based industries. The current Third Plan and Economic Transformation Programme continue to emphasize the importance of rubber manufacturing. The experience of Malaysia over a 60-year period suggests that governments of developing countries can benefit from the wealth of knowledge and technical expertise of international development agencies, such as UNIDO, to identify and prepare feasibility studies on industrial development strategies tailored to each country's economic circumstances.

Role of Rubber Manufacturing in Industrialization

In this final section the discussion presents the concluding remarks on the role the rubber products manufacturing sector has played in Malaysia's industrial development programme over the past 60 years or so. The key research question at the heart of the study is whether or not the manufacture of rubber products made from locally produced natural rubber has contributed to the country's success in moving from a primary commodity producer to an industrialized country. It is argued here that the production of natural rubber and the manufacture of rubber products have contributed in the past and continue to contribute to Malaysia's economic development.

The present day rubber products industry makes a large number of intermediate and consumer items manufactured from solid rubber and liquid latex concentrate in three main production sectors. While the latex products sector is the most important in terms of export earnings and consumption of natural rubber, the study has demonstrated that the tyre and general rubber goods sectors are substantial export earners and users of rubber raw material. The rubber manufacturing sector, therefore, is diverse and plays a significant role in the process of the diversification of Malaysia's industrial base.

The first rubber manufacturing companies were founded in the 1920s and 1930s. The majority of producers, however, were established in the period 1985 to 2005. The industry has increased in size from about 30 firms in 1950 to 350 enterprises today with some 80 per cent of businesses owned by Malaysian interests. Private investors belonging to the ethnic Chinese community control the majority of local firms. Foreign direct investment is concentrated in the industrial products and pneumatic tyre industries where 30 per cent of companies have foreign involvement. The industry's total rubber consumption of 699,000 metric tons, made up of 441,000 metric tons of natural rubber and 258,000 metric tons of synthetics, generated export revenues of RM14.5 billion. This compared with exports of 770,000 metric tons of raw rubber valued at RM7.8 billion which is half the value of rubber manufactures, thus demonstrating the industry's capacity to add value to locally produced natural rubber by the production of manufactured goods in an export-oriented industry.

A picture of a dualistic structure in the industry has emerged from the analysis of data in the MRB trade directory and a postal survey of firms in the industrial products sector. Joint venture and wholly foreign-owned manufacturers are typically larger in size, measured in terms of the amount of capital investment and employee numbers, than local businesses. Firms owned by Malaysian investors are more fully integrated into the domestic economy, compared to those owned either partially or fully by overseas interests. Moreover, local companies have stronger linkages with domestic producers and suppliers of manufacturing inputs than foreign-controlled firms. They also use the R&D facilities of the Malaysian Rubber Board as the source of manufacturing technology and technical advice, compared to firms with foreign interests, which obtain technical R&D inputs from their parent companies offshore.

The substantial presence of foreign companies is a clear indication that the steps taken by the government to encourage direct foreign investment have borne fruit. The ability of the Malaysian authorities to attract overseas manufacturers to establish production facilities and continue their operations in the country is a reflection of the congenial business environment in Malaysia. Transnational businesses have the capacity to relocate to other countries if the conditions there become more advantageous to foreign investors. However, within the industry the current composition of foreign-controlled and local investment is such that it protects the economy from any likely adverse consequences in such an eventuality. The facts that local companies control some 80 per cent of the industry, are well integrated in the domestic economy, and are ably served by local R&D institutions speak well for a stable sector in the Malaysian economy as it is firmly rooted in itself while enjoying the benefits of foreign investment.

The rubber products manufacturing sector consumes only the equivalent of 16 per cent of domestic production of natural rubber and relies on imports of latex concentrate and solid rubber from abroad for use as manufacturing inputs. It also imports a range of synthetic rubbers for use in manufacturing processes although this will become less important as the country develops its own synthetic capacity. In terms of global consumption of natural rubber, the country is ranked sixth in the world. Malaysia is the world's highest consumer of latex concentrate, using some 80 per cent of consumption to produce latex goods. The Malaysian rubber-based manufacturing industry is the largest supplier of medical examination gloves, hospital catheters, condom contraceptives and elastic latex thread to international markets. Made-in-Malaysia goods are exported to 180 countries worldwide and South-South trade is of importance to Malaysian manufacturers as well as traditional markets in industrialized countries.

The rubber goods production industry is a major employer of labour with the workforce of some 60,000 employees, accounting for 6 per cent of employment in the national manufacturing sector. For most of the 20th century, the rubber plantation industry was the largest employer of labour in the country, hiring and training three generations of managers, supervisory and clerical staff, skilled artisans and a regular and disciplined wage-earning labour force. Managers, staff and workers were thus well placed to transfer their skills to newly established manufacturing concerns as the country industrialized. The contribution of the rubber industrial sector is evident in the increase of its employment capacity from 5,700 workers in 1950 to over 60,000 employees today.

A positive factor in the development of the rubber industry is the competitive advantage that Malaysia has as the foremost authority on R&D in natural rubber. The Rubber Research Institute was established in 1925 and the MRB research centre in England began work in 1938. The R&D facilities developed over the past 90 years have made the two centres the largest research institution in the world for a single commodity and the institutes have an international reputation in the fields of rubber agronomy and polymer science. Over the years the MRB has accumulated a pool of expertise on all scientific aspects of natural rubber in the areas of plant science, agronomy, plantation management, processing of raw rubber, rubber manufacturing technology and rubber polymer science. The research work of MRB scientists and technologists greatly enhances Malaysia's R&D capacity and contributes to the knowledge economy of the nation. The strong R&D and technical support provide a competitive edge for the entire rubber production industry in the face of competition from other major producing countries.

The rubber industry has evolved from an exporter of an industrial raw material commodity to become a fully integrated industry where the expansion of downstream industries has overtaken raw rubber in importance in the export market. Implementation of the industrial development strategy in the three Industrial Master Plans gave impetus to the growth of the rubber products and rubber wood manufacturing sectors. Although Malaysia is still a significant exporter of high specification raw rubber, the rubber cultivation sector has been transformed to become a major raw material supplier to the local rubber manufacturing sector as well as supplying timber to the *Hevea* wood products industry. With these developments, the competitiveness of the rubber industry as a whole—commodity exports, rubber manufactures and *Hevea* wood products—has been greatly enhanced.

The economy of Malaysia has been transformed from one based on the production of the primary commodities, tin and rubber, to one that is classified as newly industrialized, producing a range of manufactured goods. The study argues that the process of industrial development has been built partly on the back of downstream activities arising from the *Hevea* rubber production industry. The evidence examined during the course of the study demonstrates that the manufacture of rubber products has contributed in no small measure to the success of Malaysia's industrialization programme. The rubber industry taken as a whole is an example of a vertically integrated, resource-based industry using local production of natural rubber as raw material for the manufacturing sector to produce items such as automobile tyres, latex products and general rubber goods to demanding technical standards for domestic and international markets.

Bibliography

Abdul Hamid Sawal (2001). "The Malaysian Rubber Industry: Meeting New Challenges and Moving Ahead," in *MRPMA Industry and Export Directory 2002–2003*. Subang Jaya, Malaysia: MRPMA, pp. 106–14.

Abdul Samad Hadi (1994). "Agriculture and Industry: Towards Vertical Integration," in *Transformation with Industrialization in Peninsular Malaysia*, ed. H. Brookfield. Kuala Lumpur: Oxford University Press, pp. 49–62.

Adhikari, R., C. Kirkpatrick and J. Weiss (1992). *Industrialization and Trade Policy Reform in Developing Countries*. Manchester: Manchester University Press.

Allen, G.C. and A.G. Donnithorne (1954). *Western Enterprise in Indonesia and Malaya: A Study in Economic Development*. London: Allen and Unwin.

Andaya, B.W. and L.Y. Andaya (2001). *A History of Malaysia*, 2nd ed. Basingstoke, UK: Palgrave.

Anuwar, Ali (1992). *Malaysia's Industrialization: The Quest for Technology*. Singapore: Oxford University Press.

Barlow, C. (1978). *The Natural Rubber Industry: Its Development, Technology and Economy in Malaysia*. Kuala Lumpur: Oxford University Press.

Barlow, C., ed. (1999). *Institutions and Economic Change in Southeast Asia: The Context of Development from the 1960s to 1990s*. Cheltenham, UK: Edward Elgar.

Barlow, C., S. Jayasuriya and C.S. Tan (1994). *The World Rubber Industry*. London: Routledge.

Barlow, H.S. (1996). "Epilogue," in *The RGA History of the Plantation Industry in the Malay Peninsula*, D.J.M. Tate. Kuala Lumpur: Oxford University Press, pp. 599–601.

Baulkwill, W.J. (1989). "The History of Natural Rubber Production," in *Rubber*, ed. C.C. Webster and W.J. Baulkwill. Harlow, UK: Longman, pp. 1–56.

Bowie, A. (1991). *Crossing the Industrial Divide: State, Society and the Politics of Economic Transformation in Malaysia*. New York: Columbia University Press.

Brookfield, H., ed. (1994). *Transformation with Industrialization in Peninsular Malaysia*. Kuala Lumpur: Oxford University Press.

———. (1994a). "Transformation before the late 1960s," in *Transformation with Industrialization in Peninsular Malaysia*, ed. H. Brookfield. Kuala Lumpur: Oxford University Press, pp. 1–13.

Chee, K.S. (2010). "China Compound Rubber," *Planter* 86: 531–2.

Chin, Kee Onn (1977). *Malaya Upside Down*. Kuala Lumpur: Federal Publications.

Cho, G. (1990). *The Malaysian Economy: Spatial Perspectives*. London: Routledge.

Chong, K.Y. (1982). "Trade and External Relations," in *The Political Economy of Malaysia*, ed. E.K. Fisk and H. Osman-Rani. Kuala Lumpur: Oxford University Press, pp. 184–204.

Coates, A. (1987). *The Commerce in Rubber: The First 250 Years*. Singapore: Oxford University Press.

Cooper, B.C. (2001). *Decade of Change: Malaya and the Straits Settlements, 1936–1945*. Singapore: Graham Brash.

Crouch, H. (1994). "Industrialization and Political Change," in *Transformation with Industrialization in Peninsular Malaysia*, ed. H. Brookfield. Kuala Lumpur: Oxford University Press, pp. 14–34.

Crowther, B.G., P.M. Lewis and C. Metherell (1988). "Compounding," in *Natural Rubber Science and Technology*, ed. A.D. Roberts. Oxford: Oxford University Press, pp. 177–234.

Davenport-Hines, R.P.T. and G. Jones, ed. (1989). *British Business in Asia since 1860*. Cambridge: Cambridge University Press.

Dean, W. (1987). *Brazil and the Struggle for Rubber: A Study in Environmental History*. Cambridge: Cambridge University Press.

Drabble, J. H. (1973). *Rubber in Malaya, 1876–1922: The Genesis of the Industry*. Kuala Lumpur: Oxford University Press.

———. (2000). *An Economic History of Malaysia, c. 1800–1990: The Transition to Modern Economic Growth*. Basingstoke, UK: Macmillan.

ENI (2013). Continually updated. http://www.eni.com/en_IT/home.html.

Edgar, A.T. (1958). *Manual of Rubber Planting (Malaya)*, 2nd ed. Kuala Lumpur: Incorporated Society of Planters.

Edwards, C.B. (1992). "Malaysian Industrial Policy in the 1990s," in *Industrialization and Trade Policy Reform in Developing Countries*, ed. R. Adhikari, C. Kirkpatrick and J. Weiss. Manchester: Manchester University Press, pp. 153–69.

Far Eastern Economic Review (*FEER*) (9 October 2003). Hong Kong.

Felker, G. (2001). "The Politics of Industrial Investment Policy Reform in Malaysia and Thailand," in *Southeast Asia's Industrialization: Industrial Policy, Capabilities and Sustainability*, ed. K.S. Jomo. Basingstoke, UK: Palgrave, pp. 129–82.

Fisk, E.K. and H. Osman-Rani, ed. (1982). *The Political Economy of Malaysia*. Kuala Lumpur: Oxford University Press.

Federation of Malaysian Manufacturers (FMM) (2003). *Malaysian Industries: FMM Directory, 2003*. Kuala Lumpur: FMM.

Fryer, D.W. (1970). *Emerging Southeast Asia: A Study in Growth and Stagnation.* London: George Philip and Son.

Giroud, A. (2003). *Transnational Corporations, Technology and Economic Development: Backward Linkages and Knowledge Transfer in South-east Asia.* Cheltenham, UK: Edward Elgar.

Godement, F. (1997). *The New Asian Renaissance: from Colonialism to the Post-Cold War*, tr. E.J. Parcell. London: Routledge.

Goldthorpe, C.C. (1987). "A Definition and Typology of Plantation Agriculture," *Singapore Journal of Tropical Geography* 8, 1: 26–43.

Gomez, E.T. and K.S. Jomo (1997). *Malaysia's Political Economy: Politics, Patronage and Profits.* Cambridge: Cambridge University Press.

Gopal, J. (1999). "Malaysia's Palm Oil Refining Industry," in *Industrial Technology Development in Malaysia: Industry and Firm Studies*, ed. K.S. Jomo, G. Felker and Rajah Rasiah. London: Routledge, pp. 360–95.

Graham, E. with I. Floering (1984). *The Modern Plantation in the Third World.* London: Croom Helm.

Higgins, B. (1982). "Development Planning," in *The Political Economy of Malaysia*, ed. E.K. Fisk and H. Osman-Rani. Kuala Lumpur: Oxford University Press, pp. 148–83.

International Bank for Reconstruction and Development (IBRD) (1955). *The Economic Development of Malaya.* Singapore: Government Printer.

International Rubber Research and Development Board (IRRDB) (2006). *Portrait of the Global Rubber Industry.* Kuala Lumpur: IRRDB.

_______ (2007). Continually updated. http://www.irrdb.com/irrdb/frontpage/index.php.

International Rubber Study Group (IRSG) (1990). *Planted Area Survey for Natural Rubber.* London: IRSG.

_______ (2013). Continually updated. http://www.rubberstudy.com/.

Islam, I. and A. Chowdhury (1997). *Asia-Pacific Economies: A Survey.* London: Routledge.

Jackson, J.C. (1968). *Planters and Speculators. Chinese and European Agricultural Enterprise in Malaya, 1786–1921.* Kuala Lumpur: University of Malaya Press.

Jackson, J. (2008). *The Thief at the End of the World: Rubber, Empire and the Obsessions of Henry Wickham.* London: Duckworth Overlook.

Jenkins, G.P. and A.K.K. Lai (1989). *Trade, Exchange Rate and Agricultural Policies in Malaysia.* Washington, D.C.: World Bank.

Jesudason, J.V. (1989). *Ethnicity and the Economy: The State, Chinese Business, and Multinationals in Malaysia.* Singapore: Oxford University Press.

Jomo, K.S. (1986). *A Question of Class: Capital, the State and Uneven Development in Malaya.* Singapore: Oxford University Press.

_______. (1990). *Growth and Structural Change in the Malaysian Economy.* London: Macmillan.

———, ed. (1993). *Industrialising Malaysia: Policy, Performance, Prospects*. London: Routledge.

———. (1993a). "Introduction," in *Industrialising Malaysia, Policy, Performance, Prospects*, ed. K.S. Jomo. London: Routledge, pp. 1–13.

———. (1993b). "Prospects for Malaysian Industrialisation in Light of East Asian NIC Experience," in *Industrialising Malaysia: Policy, Performance, Prospects*, ed. K.S. Jomo. London: Routledge, pp. 286–301.

———. (2001). "Rethinking the Role of Government Policy in Southeast Asia," in *Rethinking the East Asian Miracle*, ed. J.E. Stiglitz and Shahid Yusuf. New York: Oxford University Press, pp. 461–508.

———, ed. (2001a). *Southeast Asia's Industrialization: Industrial Policy, Capabilities and Sustainability*. Basingstoke, UK: Palgrave.

———, ed. (2007). *Malaysian Industrial Policy*. Singapore: NUS Press.

———. (2007a). "Industrialization and Industrial Policy in Malaysia," in *Malaysian Industrial Policy*, ed. K.S. Komo. Singapore: NUS Press, pp. 1–34.

Jomo, K.S. and C. Edwards (1993). "Malaysian Industrialisation in Historical Perspective," in *Industrialising Malaysia: Policy, Performance and Prospects*, ed. K.S. Jomo. London: Routledge, pp. 14–39.

Jomo, K.S., G. Felker and Rajah Rasiah, ed. (1999). *Industrial Technology Development in Malaysia: Industry and Firm Studies*. London: Routledge.

Jones, K.P. and P.W. Allen (1992). "Historical Developments of the World Rubber Industry," in *Natural Rubber: Biology, Cultivation and Technology*, ed. M.R. Sethuraj and N.M. Mathew. Amsterdam: Elsevier, pp. 1–25.

Junid Saham (1980). *British Industrial Investment in Malaysia, 1963–1971*. Kuala Lumpur: Oxford University Press.

Khoo, B.T. (2000). "Economic Nationalism and its Discontents: Malaysian Political Economy after July 1997," in *Politics and Markets in the Wake of the Asian Crisis*, ed. R. Robison, M. Beeson, K. Jayasuriya and H.R. Kim. London: Routledge, pp. 212–37.

Kua, Kia Soong (2007). *May 13: Declassified Documents on the Malaysian Riots of 1969*. Kuala Lumpur: Suaram Komunikasi.

Kuriakose, Baby (1992). "Primary Processing," in *Natural Rubber: Biology, Cultivation and Technology*, ed. M.R. Sethuraj and N.M. Mathew. Amsterdam: Elsevier, pp. 370–98.

Lal, D. and H. Myint (1996). *The Political Economy of Poverty, Equity and Growth: A Comparative Study*. Oxford: Clarendon Press.

Lee, Kuan Yew (1998). *The Singapore Story: Memoirs of Lee Kuan Yew*. Singapore: Singapore Press Holdings, Times Editions.

Lim, Chong-Yah (1967). *Economic Development of Modern Malaya*. Kuala Lumpur: Oxford University Press.

Lim, D., ed. (1975). *Readings on Malaysian Economic Development*. Kuala Lumpur: Oxford University Press.

———. (1975a). "The Economic, Political and Social Background to Development Planning in Malaysia," in *Readings on Malaysian Economic Development*, ed. D. Lim. Kuala Lumpur: Oxford University Press, pp. 2–10.

———. (1992). "The Dynamics of Economic Policy-Making: A Study of Malaysian Trade Policies and Performance," in *The Dynamics of Economic Policy Reform in South-East Asia and the South-West Pacific*, ed. A.J. MacIntyre and K. Jayasuriya. Singapore: Oxford University Press, pp. 94–114.

Lim, Mah Hui (1988). "Contradictions in the Development of Malay Capital: State, Accumulation and Legitimation," in *Sociology of Developing Societies: Southeast Asia*, ed. J.G. Taylor and A. Turton. London: Macmillan, pp. 19–32.

Lim, Linda Y.C. and Pang Eng Fong (1991). *Foreign Direct Investment and Industrialisation in Malaysia, Singapore, Taiwan and Thailand*. Paris: Development Centre Studies, Organisation for Economic Co-operation and Development.

Linneman, H. (1987). *Export-Oriented Industrialization in Developing Countries*. Singapore: Singapore University Press.

Lo, Sum Yee (1972). *The Development Performance of West Malaysia: 1955–1967: With Special Reference to the Industrial Sector*. Kuala Lumpur: Heinemann Educational Books (Asia).

Loadman, J. (2005). *Tears of the Tree: The Story of Rubber—A Modern Marvel*. Oxford: Oxford University Press.

Lotte Chemical Titan Holdings (2013). Continually updated. http://www.lottechem.my/.

Malaysia: Ministry of Finance. Continually updated. http://www.treasury.gov.my/.

Malaysia: Ministry of International Trade and Industry (MITI) (2010). *Third Industrial Master Plan (IMP3)*. Putra Jaya, Malaysia. Continually updated. http://www.miti.gov.my/cms/index.jsp.

Malaysia: National Economic Advisory Council (NEAC) (2009). *New Economic Model for Malaysia*. Putra Jaya, Malaysia. Continually updated. http://www.epu.gov.my/en/home.

Malaysian External Trade Development Corporation (METDC) (2013). Continually updated. http://www.matrade.gov.my/.

Malaysian Rubber Board (MRB) (2006). *Malaysian Rubber Industry and Products Directory 2006–2007*. Kuala Lumpur: MRB.

——— (2000–13). Continually updated. http://www.lgm.gov.my/.

Malaysian Rubber Export Promotion Council (MREPC) (2013). Continually updated. http://www.mrepc.com/.

Malaysian Rubber Products Manufacturers' Association (MRPMA) (2001). *Industry and Export Directory 2002–2003*. Subang Jaya, Malaysia: MRPMA, pp. 106–14.

———. (2006). *Industry and Export Directory 2006–2007*. Subang Jaya, Malaysia: MRPMA.

MacIntyre, A.J. and K. Jayasuriya, ed. (1992). *The Dynamics of Economic Policy Reform in South East-Asia and the South-West Pacific*. Singapore: Oxford University Press.

Mallet, V. (1999). *The Trouble with Tigers: The Rise and Fall of South-East Asia*. London: Harper Collins.

Md Zainuddin Salleh and Zulkifly Osman (1982). "The Economic Structure," in *The Political Economy of Malaysia*, ed. E.K. Fisk and H. Osman-Rani. Kuala Lumpur: Oxford University Press, pp. 125–47.

Mehmet, O. (1986). *Development in Malaysia: Poverty, Wealth and Trusteeship*. Beckenham, Kent: Croom Helm.

Milne, R.S. and D.K. Mauzy (1999). *Malaysian Politics under Mahathir*. London: Routledge.

Mitsubishi Corporation (2013). Continually updated. http://www.mitsubishicorp.com/jp/en/index.html.

Mohamed Ariff (1991). *The Malaysian Economy: Pacific Connections*. Singapore: Oxford University Press.

Mohamed Ariff and Muthi Semudram (1990). "Malaysia," in *Trade, Finance and Developing Countries: Strategies and Constraints in the 1990s*, ed. S. Page. Hemel Hempstead, UK: Harvester Wheatsheaf, pp. 23–55.

Morris, J.E. (1989). "Processing and Marketing," in *Rubber*, ed. C.C. Webster and W.J. Baulkwill. Harlow, UK: Longman, pp. 459–98.

Myint, H. (1972). *Southeast Asia's Economy: Development Policies in the 1970s*. Harmondsworth, UK: Penguin.

Ong, Eng Leong (2001). "Mid-term Assessment of the New Industrial Master Plan (IMP2) for the Rubber-based Sector," in *MRPMA Industry and Export Directory: 2002–2003*. Subang Jaya, Malaysia: MRPMA, pp. 115–24.

Osman-Rani, H. (1982). "Manufacturing Industries," in *The Political Economy of Malaysia*, ed. E.K. Fisk and H. Osman-Rani. Kuala Lumpur: Oxford University Press, pp. 260–86.

Overton, J. (1994). "Agriculture and Industry in Colonial Malaya," in *Transformation with Industrialization in Peninsular Malaysia*, ed. H. Brookfield. Kuala Lumpur: Oxford University Press, pp. 35–48.

Page, S., ed. (1990). *Trade, Finance and Developing Countries: Strategies and Constraints in the 1990s*. Hemel Hempstead, UK: Harvester Wheatsheaf.

Porter, M.E. (1990). *The Competitive Advantage of Nations*. London: Macmillan Press.

Pugh, Peter (1990). *Great Enterprise: A History of Harrisons & Crosfield*, ed. G. Nickalls. London: Harrisons & Crosfield.

Puthucheary, J.J. (1960). *Ownership and Control in the Malayan Economy: A Study of the Structure of Ownership and Control and Its Effects on the Development of Secondary Industries and Economic Growth in Malaya and Singapore*. Singapore: Donald Moore for Eastern Universities Press.

Rasiah, Rajah (1995). *Foreign Capital and Industrialization in Malaysia*. London: Macmillan.

________. (1999). "Government-Business Co-ordination and the Development of Eng Hardware," in *Industrial Technology Development in Malaysia: Industry and Firm Studies*, ed. K.S. Jomo, G. Felker and Rajah Rasiah. London: Routledge, pp. 231–46.

Rasiah, Rajah and K.S. Jomo (1999). "Introduction," in *Industrial Technology Development in Malaysia: Industry and Firm Studies*, ed. K.S. Jomo, G. Felker and Rajah Rasiah. London: Routledge, pp. 1–20.

Roberts, A.D., ed. (1988). *Natural Rubber Science and Technology*. Oxford: Oxford University Press.

Robison, R., M. Beeson, K. Jayasuriya and H.-R. Kim, ed. (2000). *Politics and Markets in the Wake of the Asian Crisis*. London: Routledge.

Rock, M.T. and M. Sheridan (2007). "The Political Economy of Industrial Policy in Malaysia," in *Malaysian Industrial Policy*, ed. K.S. Jomo. Singapore: NUS Press, pp. 179–221.

Schätzl, L.H., ed. (1988). *Growth and Spatial Equity in West Malaysia*. Singapore: Institute of Southeast Asian Studies.

________. (1988a). "Economic Development and Economic Policy," in *Growth and Spatial Equity in West Malaysia*. Singapore: Institute of Southeast Asian Studies.

Sethuraj, M.R. and N.M. Mathew, ed. (1992). *Natural Rubber: Biology, Cultivation and Technology*. Amsterdam: Elsevier.

Shamsul Bahrin (Tunku) and P.D.A. Perera (1977). *FELDA: 21 Years of Land Development*. Kuala Lumpur: Federal Land Development Authority.

Shamsul Bahrin (Tunku) and Lee Boon Thong (1988). *FELDA: 3 Decades of Evolution*. Kuala Lumpur: Federal Land Development Authority.

Shepherd, G. (1980). "Policies to Promote Industrial Development," in *Malaysia: Growth and Equity in a Multiracial Society*, ed. K. Young, W.C.F. Bussink and P. Hasan. Baltimore: John Hopkins University Press, pp. 182–210.

Silcock, T.H. and E.K. Fisk, ed. (1963). *The Political Economy of Independent Malaya: A Case-Study in Development*. Canberra: Australian National University.

Snodgrass, D.R. (1980). *Inequality and Economic Development in Malaysia*. Kuala Lumpur: Oxford University Press.

Stiglitz, J.E. (2002). *Globalization and Its Discontents*. London: Allen Lane.

Stiglitz, J.E. and Shahid Yusuf (2001). *Rethinking the East Asian Miracle*. New York: Oxford University Press.

Synthomer (2013). Continually updated. http://www.synthomer.com/.

Tate, D.J.M. (1996). *The RGA History of the Plantation Industry in the Malay Peninsula*. Kuala Lumpur: Oxford University Press.

Taylor, J.G. and A. Turton (1988). *Sociology of Developing Societies: Southeast Asia*. London: Macmillan.

Taylor, M. and M. Ward (1994). "Industrial Transformation since 1970: The Context and the Means," in *Transformation with Industrialization in Peninsular Malaysia*, ed. H. Brookfield. Kuala Lumpur: Oxford University Press, pp. 95–121.

________. (1994a). "Sectoral Dimensions of Regional Change," in *Transformation with Industrialization in Peninsular Malaysia*, ed. H. Brookfield. Kuala Lumpur: Oxford University Press, pp. 149–68.

Tham, S.-Y. and Mahani Zainal-Abidin (1999). "Industrial Institutions: The Case of Malaysia," in *Institutions and Economic Change in Southeast Asia: The Context of Development from the 1960s to 1990s*, ed. C. Barlow. Cheltenham, UK: Edward Elgar, pp. 55–71.

Thoburn, J.T. (1973/75). "Exports and Economic Growth in West Malaysia," in *Readings in Malaysian Economic Development*, ed. D. Lim. Kuala Lumpur: Oxford University Press, pp. 12–27. (First published in *Oxford Bulletin of Economics and Statistics* 35, 2.)

________. (1973/75a). "Exports and the Malaysian Engineering Industry," in *Readings in Malaysian Economic Development*, ed. D. Lim. Kuala Lumpur: Oxford University Press, pp. 28–46. (First published in *Oxford Economic Papers* 25: 88–111.)

________. (1977). *Primary Commodity Exports and Economic Development: Theory, Evidence and a Study of Malaysia*. London: John Wiley & Sons.

Tun Abdul Razak Research Centre (TARRC) (2007–13). Continually updated. http://www.tarrc.co.uk/.

United Nations Conference on Trade and Development (UNCTAD) (1996). *East Asian Development: Lessons for a New Global Environment, Study No. 4, Lessons from Growth and Structural Change in Second-tier South East Asian Newly Industrializing Countries*, K.S. Jomo. Geneva: United Nations.

________ (1998). *The Export Manufacturing Experience of Indonesia, Malaysia and Thailand: Lessons for Africa, Discussion Paper No. 137*, Rajah Rasiah. Geneva: United Nations.

United Nations Industrial Development Organization (UNIDO) (1985). *Medium and Long Term Industrial Master Plan, Malaysia, 1986–95, Executive Highlights*, Kuala Lumpur: UNDP/UNIDO. Cited in M. Taylor and M. Ward (1994). "Industrial Transformation since 1970: Context and Means," in *Transformation with Industrialization in Peninsular Malaysia*, ed. H. Brookfield. Kuala Lumpur: Oxford University Press, pp. 95–121.

________. (1991). *Malaysia: Sustaining the Industrial Investment Momentum*. Vienna: United Nations Industrial Development Organization.

Van Helten, J-J. and G. Jones (1989). "British Business in Malaysia and Singapore since the 1870s," in *British Business in Asia since 1860*, ed. R.P.T. Davenport-Hines and G. Jones. Cambridge: Cambridge University Press, pp. 157–88.

Verbruggen, H. (1987). "The Case of Malaysia," in *Export-Oriented Industrialization in Developing Countries*, ed. H. Linneman. Singapore: Singapore University Press, pp. 363–80.

Voon, P.K. (1976). *Western Rubber Planting Enterprise in Southeast Asia, 1876–1921*. Kuala Lumpur: University of Malaya Press.

Webster, C.C. and W.J. Baulkwill, ed. (1989). *Rubber*. Harlow, UK: Longman.

Wheelwright, E.L. (1963). "Industrialization in Malaya," in *The Political Economy of Independent Malaya: A Case-Study in Development*, ed. T.H. Silcock and E.K. Fisk. Canberra: Australian National University, pp. 210–41.

———. (1965). *Industrialization in Malaysia*. Melbourne: Melbourne University Press.

Wong, J. (1979). *ASEAN Economics in Perspective*. London: Macmillan.

World Bank (1993). *The East Asian Miracle: Economic Growth and Public Policy*. New York: Oxford University Press.

Young, K. (1980). "The New Economic Policy and Long Term Development Issues," in *Malaysia: Growth and Equity in a Multiracial Society*, K. Young, W.C.F. Bussink and P. Hasan. Baltimore: Johns Hopkins University Press.

Young, K., W.C.F. Bussink and P. Hasan (1980). *Malaysia: Growth and Equity in a Multiracial Society*. Baltimore: Johns Hopkins University Press.

Index